Non-isotopic Methods in Molecular Biology

The Practical Approach Series

SERIES EDITORS

D. RICKWOOD
Department of Biology, University of Essex
Wivenhoe Park, Colchester, Essex CO4 3SQ, UK

B. D. HAMES
Department of Biochemistry and Molecular Biology
University of Leeds, Leeds LS2 9JT, UK

Affinity Chromatography
Anaerobic Microbiology
Animal Cell Culture
  (2nd Edition)
Animal Virus Pathogenesis
Antibodies I and II
Basic Cell Culture
Behavioural Neuroscience
Biochemical Toxicology
Biological Data Analysis
Biological Membranes
Biomechanics — Materials
Biomechanics — Structures and
  Systems
Biosensors
Carbohydrate Analysis
  (2nd Edition)
Cell–Cell Interactions
The Cell Cycle
Cell Growth and Division
Cellular Calcium
Cellular Interactions in
  Development
Cellular Neurobiology

Centrifugation (2nd Edition)
Clinical Immunology
Computers in Microbiology
Crystallization of Nucleic Acids
  and Proteins
Cytokines
The Cytoskeleton
Diagnostic Molecular Pathology
  I and II
Directed Mutagenesis
DNA Cloning: Core Techniques
DNA Cloning: Expression
  Systems
Drosophila
Electron Microscopy in Biology
Electron Microscopy in
  Molecular Biology
Electrophysiology
Enzyme Assays
Essential Developmental
  Biology
Essential Molecular Biology I
  and II
Experimental Neuroanatomy

Non-isotopic Methods in Molecular Biology

A Practical Approach

Edited by

E. R. LEVY

Oxford Medical Genetics Laboratories,
Churchill Hospital, Oxford

and

C. S. HERRINGTON

Nuffield Department of Pathology and Bacteriology,
John Radcliffe Hospital, Oxford

————at————
OXFORD UNIVERSITY PRESS
Oxford New York Tokyo

This book has been printed digitally in order to ensure its continuing availability

OXFORD
UNIVERSITY PRESS

Great Clarendon Street, Oxford OX2 6DP

Oxford University Press is a department of the University of Oxford.
It furthers the University's objective of excellence in research, scholarship,
and education by publishing worldwide in

Oxford New York

Auckland Bangkok Buenos Aires Cape Town Chennai
Dar es Salaam Delhi Hong Kong Istanbul Karachi Kolkata
Kuala Lumpur Madrid Melbourne Mexico City Mumbai Nairobi
São Paulo Shanghai Singapore Taipei Tokyo Toronto
with an associated company in Berlin

Oxford is a registered trade mark of Oxford University Press
in the UK and in certain other countries

Published in the United States
by Oxford University Press Inc., New York

© Oxford University Press, 1995

A catalogue record for this book is available from the British Library

Library of Congress Cataloging in Publication Data
Non-isotopic methods in molecular biology : a practical approach /
edited by E. R. Levy and C. S. Herrington.
(Practical approach series)
Includes bibliographical references and index.
1. Nucleic acid probes. I. Levy, E. R. II. Herrington, C. S.
QP620.N66 1995 574.87'328'028—dc20 94-40129

ISBN 0-19-963456-4 (Hbk)
ISBN 0-19-963455-6 (Pbk)

Preface

There is increasing awareness of the use and inherent advantages of non-isotopic methods for the detection and analysis of nucleic acids. In many situations, these have supplanted those requiring radioactive materials as they are generally cheaper, need less equipment, and are safer. In this volume, protocols used routinely in both diagnostic and research laboratories are presented to allow the experienced laboratory worker to convert existing techniques using radioactive probes to those using non-isotopic probes, and to enable the beginner with no previous experience of radiolabelling to start with non-isotopic techniques. These protocols are presented alongside trouble-shooting guides where applicable and details of commercial kits where available.

Chapter 1 describes the different non-isotopic analogues that can be used to replace radioactive isotopes in well-established labelling techniques and includes detailed protocols for their use. The chapter also includes more recent techniques such as primed *in situ* labelling (PRINS).

Chapters 2 to 5 contain detailed protocols using the probes produced in Chapter 1 for the *in situ* analysis of DNA and RNA in various systems from whole chromosomes to intact cells and tissues. Chapter 2 deals specifically with cytogenetic analysis, both metaphase and interphase, together with a review of the equipment available for this type of study. Chapters 3 to 5 describe *in situ* hybridization techniques applicable to the detection of DNA and RNA in both intact cells and tissues, including the use of synthetic oligonucleotides for RNA *in situ*. Protocols for the combination of PCR amplification and *in situ* hybridization (so-called PCR *in situ* hybridization) are included as are detailed guidelines for the combination of immunohistochemical techniques with the non-isotopic detection of nucleic acids.

Filter hybridization assays are described in detail in Chapter 6, which includes protocols for colourimetric and chemiluminescent detection methods.

Chapters 7 and 8 introduce techniques for DNA and RNA analysis which avoid the use of a hybridization assay, relying instead on the specificity of the polymerase chain reaction. Chapter 7 describes PCR amplification of multi-allelic systems for genomic DNA analysis, while Chapter 8 concentrates on the amplification and detection of RNA following initial conversion to complementary DNA (reverse transcriptase PCR).

The text is designed to provide an introduction to non-isotopic techniques applicable using relatively simple equipment. We thank Oxford University Press and the individual contributors for their efforts in this regard.

Oxford
March 1995

E.R.L.
C.S.H.

Contents

Contents

Contents

Contents

7. Non-isotopic DNA analysis   183

Y.-M. Dennis Lo and Wajahat Z. Mehal

Contents

Contributors

M. S. BASHIR
Department of Pathology, University of Leeds, Leeds LS2 9JT, UK.

G. BROWNE
Department of Pathology, University of Leeds, Leeds LS2 9JT, UK.

M. CROWLEY
University Department of Pathology, University College Cork, Ireland.

C. T. DOYLE
University Department of Pathology, University College Cork, Ireland.

M. F. EVANS
Nuffield Department of Pathology and Bacteriology, Level 4 Academic Block, John Radcliffe Hospital, Oxford OX3 9DU, UK.

I. HEALEY
University Department of Pathology, University College Cork, Ireland.

C. S. HERRINGTON
Nuffield Department of Pathology and Bacteriology, Level 4 Academic Block, John Radcliffe Hospital, Oxford OX3 9DU, UK.

A. H. N. HOPMAN
Department of Molecular Cell Biology, University of Limburg, PO Box 616, 6200MD, Maastricht, The Netherlands.

A. HORSTMAN
Department of Pathology, Free University Hospital, De Boelelaan 1117, 1081 HV, Amsterdam, The Netherlands.

J. R. HUGHES
MRC Molecular Haematology Unit, Institute of Molecular Medicine, John Radcliffe Hospital, Oxford OX3 9DU, UK.

N. M. JIWA
Department of Pathology, Free University Hospital, De Boelelaan 1117, 1081 HV, Amsterdam, The Netherlands.

K. W. KLINGER
Integrated Genetics Inc., PO Box 9322, One Mountain Road, Framington, MA 01701, USA.

R. J. LANDERS
University Department of Pathology, University College Cork, Ireland.

Contributors

E. R. LEVY
Oxford Medical Genetics Laboratories, Oxford Radcliffe Hospital, The Churchill, Headington, Oxford OX3 7LJ, UK.

F. A. LEWIS
Department of Pathology, University of Leeds, Leeds LS2 9JT, UK.

Y.-M. DENNIS LO
Department of Haematology, Level 4 Academic Block, John Radcliffe Hospital, Oxford OX3 9DU, UK.

WAJAART Z. MEHAL
Nuffield Department of Pathology and Bacteriology, Level 4 Academic Block, John Radcliffe Hospital, Oxford OX3 9DU, UK.

C. J. L. M. MEIJER
Department of Pathology, Free University Hospital, De Boelelaan 1117, 1081 HV, Amsterdam, The Netherlands.

H. MULLINK
Department of Pathology, Free University Hospital, De Boelelaan 1117, 1081 HV, Amsterdam, The Netherlands.

J. J. O'LEARY
Nuffield Department of Pathology and Bacteriology, Level 4 Academic Block, John Radcliffe Hospital, Oxford OX3 9DU, UK.

JAMES HOWARD PRINGLE
Department of Pathology, University of Leicester, Clinical Sciences Building, Leicester Royal Infirmary, Leicester LE2 7LX, UK.

F. C. S. RAMAEKERS
Department of Molecular Cell Biology, University of Limburg, PO Box 616, 6200MD, Maastricht, The Netherlands.

E. RIEGER
Department of Pathology, Free University Hospital, De Boelelaan 1117, 1081 HV, Amsterdam, The Netherlands.

E. J. M. SPEEL
Department of Molecular Cell Biology, University of Limburg, PO Box 616, 6200MD, Maastricht, The Netherlands.

EIICHI TAHARA
Department of Pathology, Hiroshima University School of Medicine, 1-2-3 Kasumi, Minami-ku, Hiroshima 734, Japan.

C. E. M. VOORTER
Department of Molecular Cell Biology, University of Limburg, PO Box 616, 6200MD, Maastricht, The Netherlands.

Contributors

W. VOS
Department of Pathology, Free University Hospital, De Boelelaan 1117, 1081 HV, Amsterdam, The Netherlands.

HIROSHI YOKOZAKI
Department of Pathology, Hiroshima University School of Medicine, 1-2-3 Kasumi, Minami-ku, Hiroshima 734, Japan.

Colour plates

Abbreviations

AAF	*N*-acetoxy-*N*-2-acetylaminofluorene
ABC-PO	avidin–biotin–peroxidase complex
AEC	3-amino-9-ethyl carbazole
AMCA	aminomethylcoumarine
AMPPD	3-(4-methoxyspiro{1,2-dioxetane-3,2′-tricyclo[3.3.1.1$^{3.7}$]-decan}4-yl)phenyl phosphate
AP	alkaline phosphatase
APAAP	alkaline phosphatase–anti-alkaline-phosphatase
APES	3-aminopropyltriethoxysilane
ATCC	American Type Culture Collection
BAC	bacterial artificial chromosome
BCIP	bromo-chloro-indolyl phosphate
BL	bioluminescence
BSA	bovine serum albumin
CGH	comparative genomic hybridization
CISS	chromosomal *in situ* suppression
CL	chemiluminescence
CsCl	caesium chloride
DAB	diaminobenzidine
DABCO	1,4-diazobicyclo[-2.2-] octane
DAPI	4,6 diamino-2-phenyl-indole
DEPC	diethylpyrocarbonate
DIG	digoxigenin
DISH	DNA *in situ* hybridization
DNP	dinitrophenol
DOP	degenerate oligonucleotide primer
ECL	enhanced chemiluminescence
FISH	fluorescent *in situ* hybridization
FITC	fluorescein isothiocyanate
GTC	guanidinium isothiocyanate
HPV	human papillomavirus
HRP	horseradish peroxidase
IAP	immuno-alkaline phosphatase
IBG	immuno-beta galactosidase
IC	immunocytochemistry
IGSS	immunogold–silver staining
IMS	industrial methylated spirits
IPO	immunoperoxidase
IRS	interspersed repetitive sequences

Abbreviations

LAM	luminescence amplifying materials
MoPS	3-N-morpholino-propanesulfonic acid
NBT	nitroblue tetrazolium
NISH	non-isotopic *in situ* hybridization
OCT	optimal cutting temperature
PAP	peroxidase–anti-peroxidase
PBS	phosphate buffered saline
PCR	polymerase chain reaction
PHA	phytohaemagglutinin
PLL	poly-L-lysine
PVP	polyvinylpyrrolidone
RISH	RNA *in situ* hybridization
SSC	standard saline citrate
TMB	tetramethylbenzidine
TRITC	tetramethyl rhodamine isothiocyanate
TBS	Tris buffered saline
RT	reverse transcription
YAC	yeast artificial chromosome

1

Probe labelling methods

A. H. N. HOPMAN, E. J. M. SPEEL, C. E. M. VOORTER, and
F. C. S. RAMAEKERS

1. Introduction

Several methods have been developed to detect specific nucleic acid sequences in various preparations, such as *in situ* localized DNA or RNA in intact cells or biochemically isolated DNA or RNA immobilized on membranes. In the latter case nucleic acids are either directly immobilized on filters (spot hybridization) or transferred to the filter after electrophoretic fractionation. The presence of specific nucleic acids is then detected on the membrane by means of hybridization with labelled nucleic acid probes. Techniques aimed at the localization of target DNA or RNA sequences within microscopic preparations, so called *in situ* hybridization, require several pre-treatment steps to allow the penetration of such probes into the matrix of the cell. Nucleic acid probes can be labelled either radioactively, followed by autoradiographic detection, or non-radioactively, enabling direct detection by different forms of microscopy.

Probes for (non)-radioactive *in situ* or filter hybridizations can be obtained by cloning specific nucleic acid sequences in plasmids, cosmids, or single- or double-stranded phage vectors. Additionally, chemically synthesized short oligonucleotide probes can be used. These synthetic oligonucleotides, often derived from a known DNA/RNA/amino acid sequence, obviate the requirements for cloning mRNA or genomic DNA sequences to obtain gene-specific probes. Therefore, they have become useful tools in biomedical, veterinary, and agricultural research and in diagnosis.

During the past 10 years non-radioactive hybridization techniques have gained ground for obvious reasons. The inconveniences of radioisotopes are avoided, they enable sharp topological localization (for *in situ* hybridization), and allow the use of multitarget detection protocols. The preparations can be screened immediately since autoradiography is not required. In this chapter, non-radioactive labelling and detection techniques will be discussed with special emphasis on the choice of labels and their respective immunochemical- or affinity-detection systems.

Several approaches have been described for nucleic acid labelling in non-radioactive hybridization protocols. The marker molecules can be linked to different sites of the nucleic acid probe (see *Figure 1*), but to obtain specific hybridization between modified probe and target DNA/RNA it is important that the modification procedure does not influence the structure of the probe to such an extent that annealing is inefficient. Furthermore, the label must be firmly attached to the nucleic acid probe in order to withstand different fixation and denaturation steps. Where the hapten is the marker group, the molecule should be sufficiently exposed to enable high-affinity (antibody) binding necessary for high sensitivity, and where the marker group is an enzyme, enzymatic activity should be retained during its coupling to the probe and under the hybridization conditions.

Two major approaches of probe labelling can be distinguished, (a) labelling by chemical modification (Section 2) and (b) labelling by enzymatic modification (Section 3).

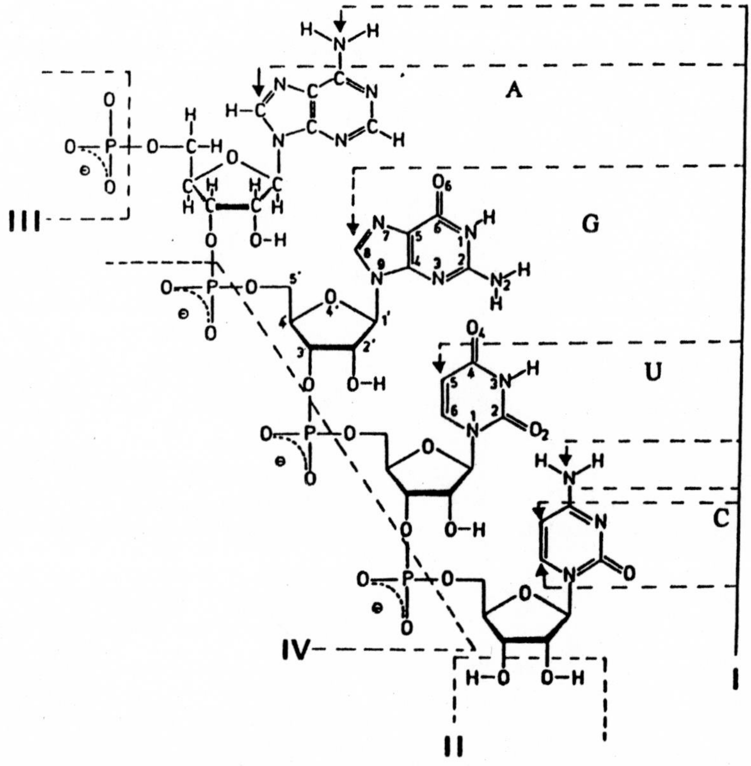

Figure 1. A simplified model of a nucleic acid chain showing sites at which modification is possible. I, individual nucleotides; II, the 3'-terminus (RNA); III, the 5'-terminal phosphate group; IV, by ionic interactions between the phosphate groups and, for example, positively charged proteins enabling cross-linking to DNA.

2. Chemical labelling of nucleic acid probes (see *Table 1*)

2.1 RNA labelling at the 3'-terminus

Selective periodate oxidation of the 3'-terminal *cis*-diol of RNAs results in the formation of a reactive dialdehyde, which reacts readily with nucleophilic amino compounds, such as amines, semicarbazides, thiosemicarbazides, hydrazines, and hydrazides. Fluorescent thiosemicarbazides couple efficiently to oxidized RNAs (1). The thiosemicarbazides are prepared by reaction of the isothiocyanates of the commonly used fluorochromes such as fluorescein (FITC), and tetramethylrhodamine (TRITC) with hydrazine. The resulting fluorescent RNA probes are stable during *in situ* hybridization reactions, provided that the hybridization temperatures used are not too high. The end-products can be visualized directly or indirectly, with increased sensitivity, using the relevant antibody.

Modifications at the terminal ends of the polynucleotide chain, in general, lead to undistorted reannealing. Relatively large molecules can be coupled to nucleic acid probes without any effect on the hybridization efficiency, provided 75–150 nucleotides remain free to reanneal with the complementary strand.

Table 1. Labels used for chemical labelling of nucleic acid probes

Label	Probe	Site of interaction	Reference
FITC/TRITC	RNA	3'-terminus	1
Dinitrophenyl	DNA/RNA	randomly	2
Acetylaminofluorene	DNA/RNA	guanidine	3, 4
Biotin	DNA/RNA	randomly	5
Mercury	DNA/RNA	cytidine/uracil	6
Aminogroup	DNA/RNA	pyrimidine	9
Sulfone	DNA/RNA	pyrimidine	8
Enzymes	DNA/RNA	randomly	10

FITC, fluorescein; TRITC, tetramethylrhodamine

2.2 Dinitrophenyl labelling

Dinitrophenyl groups can be coupled to nucleic acid sequences by the reaction of 2,4-dinitrophenylbenzaldehyde with a DNA probe at pH 12 (2). In this way about 5% of the nucleotides in any given sequence can be labelled with the dinitrophenyl group, with no significant reduction of the hybridization efficiency or specificity. Anti-dinitrophenyl antibodies are used for the immunocytochemical detection of the *in situ* hybridized nucleic acid probes.

2.3 Acetylaminofluorene labelling

Several workers have described the chemical modification of guanine residues in RNA and DNA probes using the carcinogenic compound *N*-acetoxy-*N*-2-acetylaminofluorene (AAF) (3, 4). The fluorene derivative is covalently linked, mainly to the C-8 position of the guanine residues. About 5% of the bases in a given nucleic acid sequence can be modified with a slight decrease (about 5%) of the melting temperature (T_m) of the hybrid to be formed. These AAF-modified probes can be used successfully in combination with anti-AAF antibodies for the detection of DNA sequences immobilized on nitrocellulose filters and for the detection of DNA target molecules in *in situ* hybridization.

2.4 Photoactivation for labelling

A synthetic photoreactive analogue of biotin can be randomly introduced into nucleic acid sequences (5). The biotin molecule is linked to an arylamide which can be photoactivated to generate highly reactive aryl nitrenes that couple to the aromatic bases of nucleic acids. The linker contains a positively charged tertiary amino group for phosphate group interaction. A high local concentration of the reagent can thus be reached in the vicinity of the nucleic acid which favours photo cross-linking. Different photoreactive analogues are nowadays commercially available and can be used to label DNA, RNA probes, or proteins (Boehringer–Mannheim, BRL, Pearce). This technique is useful for generating full length probes, stable under a variety of conditions such as alkaline pH, high temperature, and irradiation.

2.5 Mercuration of nucleic acids

Covalent mercuration of nucleic acids (modification site cytidine, uridine) enables the introduction of mercury atoms into probes. These can serve as an anchoring point for hapten-carrying mercaptans after hybridization to the target DNA (6). The sulfhydryl-hapten-carrying ligand is stably bound to the probe; trinitrophenyl, biotin, fluorescein, and tetramethylrhodamine have been used as haptens.

2.6 Bisulfite modifications

Cytidine residues can be modified using bisulfite, resulting in the addition of sodium bisulfite to the 5,6-double bond of the pyrimidine base. This addition facilitates the substitution of the exocyclic amino group by transamination with an appropriate amine (7). Two detection systems based on this bisulfite interaction are now commercially available. One introduces a sulfone group on to the probe after the bisulfite reaction, the modified probe then detected with antibodies raised to sulfonated nucleic acids (FMC Bioproducts; Chemiprobe) (8). Alternatively after modification using bisulfite, diamines can be coupled, resulting in amino-group-labelled cytidines. Fluorescent labels or *N*-

hydroxysuccinimide esters can be coupled to these amino groups (9). A hapten modification procedure is commercially available (Boehringer–Mannheim) in which the cytidine residues are derivatized (resulting in a primary amine) and the modified bases then coupled to *N*-hydroxysuccinimide activated digoxigenin. Thereafter immunochemical detection is performed with anti-digoxigenin antibodies.

2.7 Enzymes

Different chemical approaches have been published describing enzyme labelling of probes. Some protocols are based on the ionic interaction of positively charged macromolecules with the phosphate groups of the nucleic acid probe, followed by cross-linking to the nitrogen base residues. Peroxidase and alkaline phosphatase have been coupled as marker molecules directly to a single-stranded DNA probe (10). To enable covalent glutaraldehyde coupling of these uncharged enzymes to the probes, a small polyanionic molecule (in this case polyethyleneimine using *p*-benzoquinone) first has to be coupled to the enzyme.

The thermal stability of hybrids formed between the target and the enzyme–nucleic acid probe is only slightly reduced as a result of this modification procedure. This is perhaps rather surprising, considering the fact that enzyme protein is also present in the region containing the specific DNA insert. Complexes with molecular weights up to 1×10^7 daltons were used for hybridization on to nitrocellulose and the number of enzyme proteins reached levels of 1 molecule per 7.4 nucleotides, implying mass ratios of protein/nucleic acid from 1:1 (11) to 25:1 (12). Although the substitution level is high, there are apparently sufficient base pairs free for hybridization to the target sequence.

2.8 Chemical modification of oligonucleotides

Oligonucleotides can be labelled selectively at the 5′-terminus through a linker arm with a chemically reactive group (13, 14). The terminal phosphate group is allowed to react with a water-soluble carbodiimide in imidazole buffer to give a 5′-phosphorimidazolide. Exposure of the activated compound to amine-containing molecules in aqueous solution results in the production of a 5′-labelled probe and a biotin moiety can then be coupled to amine-containing oligonucleotides. These probes hybridize effectively to target nucleic acids on filters. When a phosphate group is absent, a 5′-phosphate group can be coupled using polynucleotide kinase with ATP as a substrate; alternatively other amines such as poly-L-lysine and protein bovine serum albumin can be used. Biotin moieties can also be coupled to the 5′-end of synthetic oligonucleotides (15). This involves chemical coupling of a uridyl residue followed by a strategy similar to coupling to the vicinal OH-groups at the 3′-terminus. The residue is periodate-oxidized, reacted with biotin hydrazide, and stabilized by reduction with sodium borohydride. Oligo-nucleotides containing free sulfhydryl groups to which thiol-specific molecules are subsequently attached can be synthesized following this protocol (16).

In summary, chemical modification (reviewed in refs 17–20) allows large quantities of nucleic acid probe to be modified and these chemical reactions, when well standardized, are highly reproducible. Modification of the nucleotides is random through the entire nucleic acid probe. A possible drawback of chemical labelling is the instability of reagents used for the nucleic acid modification procedure. Furthermore, the exact location in the sequence is not always known and some chemical reactions are complex. For most of the chemical labelling approaches no kits are available commercially. For these reasons enzymatic modification of nucleic acid probes is nowadays the method of choice for most applications.

3. Enzymatic labelling of nucleic acid probes (see *Table 2*)

3.1 Choice of label

3.1.1 Biotin

The enzymatic synthesis of haptenized polynucleotides as hybridization probes was first described (21) using analogues of dUTP and UTP which contained a biotin molecule, covalently bound to the C-5 position of the pyrimidine ring through an allylamine linker arm. These nucleotides were found to be efficient substrates for several polymerases *in vitro*. Synthesized polynucleotides, containing low levels of conjugated biotin (5% of all bases), showed similar hybridization characteristics to unsubstituted probes. This enables immuno-logical detection of probe–target DNA hybrids using anti-biotin antibodies. The approach of substitution at the C-5 position of the base, allowing the introduction of smaller groups without reducing the polymerase and hybridization reactions significantly, was first reported by Dale *et al.* (22). They found that mercury substitution at the C-5 position of the pyrimidine uridine did not affect the polymerase or hybridization reaction. The method to introduce an affinity label at this site of the polynucleotide was probably inspired by these results (21).

Biotin nucleotides with longer spacer arms were synthesized to improve the efficiency of the interaction of biotin with avidin (23, 24).

The use of biotin-labelled polynucleotides triggered other investigators to modify nucleotide triphosphates or nucleotides at other sites in the pyrimidine or purine ring. These analogues could be incorporated into DNA probes by different enzymes (see *Table 2*). By enzymatic labelling the exact location of the modification in the DNA probe is known, and, therefore, selective modification is possible (see Section 3.2). Different types of nucleic acid probes require different labelling protocols and enzymes. However, for the different labelled triphosphates the same protocols can be used (see Section 3.2 and *Table 2*). Additionally, amino groups can be enzymatically incorporated (Section 3.1.5), after which marker molecules can be introduced. Using this approach haptens that would normally not be incorporated because of their large size, can be incorporated directly.

3.1.2 Dinitrophenyl

Dinitrophenyl analogues of 8-aminohexyl adenosine 5'-triphosphate have been synthesized (25). These derivatives can be used as substrates for transferase and polymerase reactions, enabling enzymatic synthesis of dinitrophenyl group-containing DNA probes (Section 3.2 and *Table 2*). The probe–DNA hybrids are detected using anti-dinitrophenyl antibodies.

3.1.3 Digoxigenin

Digoxigenin (Boehringer–Mannheim) is linked to uridine nucleotides at the 5-position of the pyrimidine ring via a C-11 spacer arm (26), and is incorporated into the probes by the techniques described in Section 3.2 and *Table 2*.

3.1.4 Fluorochromes

The first available fluorochromized nucleotides were fluorescein-modified dideoxynucleotides which were produced for non-radioactive DNA sequencing. Shortly thereafter fluorescein-, rhodamine-, and coumarin-modified dUTPs were synthesized. Fluorochrome-labelled nucleotides can be enzymatically incorporated into probes (see Section 3.2 and *Table 2*) and detected directly or indirectly by specific antibodies, thereby increasing the sensitivity of the reaction. Different fluorochrome-labelled triphosphates are commercially available (Amersham International).

3.1.5 Amino-group

Several C-5 substituted pyrimidine nucleotide triphosphates have been described as substrates for DNA polymerases. Amino-group substituted DNA (Sigma, Clontech, Enzo) can be prepared by the techniques described in Section

Table 2. Enzymatic labelling of probes

Label	Enzymes	Probe
X-dUTP: X = biotin; digoxigenin; dinitrophenyl; FITC; AMCA; TRITC; **X-dATP**: X = biotin; aminohexyl	DNA polymerase 1 Klenow enzyme	ds DNA
X-UTP: X = biotin; digoxigenin; dinitrophenyl; FITC; AMCA; TRITC; **X-ATP**: X = biotin; aminohexyl	SP6, T3, and T7 RNA polymerase	ss RNA
X-ddUTP or dUTP: X = digoxigenin; biotin; FITC; aminohexyl	terminal transferase	Oligonucleotides 3'-end

FITC, fluorescein; TRITC, tetramethylrhodamine; AMCA, aminomethylcoumarin

3.2 using amino-(spacer)-dUTP as substrate. Several of these amino-dUTPs, like 5-aminoallyl-dUTP (21) or aminohexyl-ATP are commercially available. Once incorporated into DNA, the amines may serve as sites for attachment of, for example, biotin, haptens, fluorochromes, or enzymes. This approach enables the introduction of those marker molecules which, when directly coupled to dUTP, would make the dUTP analogue unacceptable as a substrate.

3.2 Choice of labelling method

Several approaches for enzymatic non-radioactive labelling of nucleic acid probes have been described. For DNA probes, labelling can be achieved either by nick-translation of double-stranded (ds) DNA (*Protocol 1*; 21, 27) or by random priming of linearized, single-stranded (ss) DNA (*Protocol 2*; 28). If sequences are cloned in phage M13, containing ssDNA, labelling can be performed by primer extension using a DNA polymerase I, that lacks exonuclease activity (Klenow fragment). If the DNA probe is generated by polymerase chain reaction (PCR), labelling may also be performed by means of PCR in a second round of amplification, using similar conditions, but including a labelled nucleotide (*Protocol 3*; 29–33). A specialized version of this procedure is the primed *in situ* (PRINS) labelling of DNA (*Protocol 4*; 34–36). Single-stranded RNA probes are generated by *in vitro* transcription, using plasmids, which contain bacteriophage RNA polymerase-specific promoters (*Protocol 5*; 37). Labelling of synthetic oligonucleotides is mainly performed by 3'-end labelling using terminal deoxynucleotidyl transferase (*Protocol 6*; 38–41), although other methods have been described.

3.2.1 Nick-translation

Nick-translation (*Protocol 1*) is most widely used for non-radioactive labelling of DNA probes. Several labelled nucleotides as well as several kits for nick-translation (for example Gibco–BRL, Boehringer–Mannheim, Amersham International) are commercially available.

Protocol 1. Nick-translation

Equipment and reagents

- DNA
- 10 × nick-translation buffer (10 × NT): 0.5 M Tris–HCl, pH 7.8, 50 mM MgCl$_2$, 0.5 mg/ml BSA (nuclease free)
- 10 × deoxynucleotide triphosphate (10 × dNTPs): 0.5 mM dATP, dGTP, dCTP, dTTP with one triphosphate replaced by the labelled nucleotide [a]
- 100 mM dithiothreitol (DTT)
- labelled triphosphate: e.g. digoxigenin-11-dUTP (DIG-dUTP) (Boehringer–Mannheim), biotin-16-dUTP (bio-dUTP) (Sigma), fluorescein-12-dUTP (FITC-dUTP)
- (Boehringer–Mannheim), biotin-7-dATP (bio-dATP) [a] (Amersham International, UK)
- DNase I (1 mg/ml)
- DNA polymerase I (5 U/μl)
- glycogen (20 mg/ml)
- 0.5 M EDTA, pH 8.0
- 3 M sodium acetate, pH 5.6
- Sephadex G 50, or spun columns
- carrier herring sperm DNA
- 10 mM Tris–HCl, pH 8.0, 1mM EDTA
- ethanol (−20°C)

Method

1. Mix the following in a microcentrifuge tube:
   - DNA   1 μg
   - 10 × NT buffer   5 μl
   - 10 × dNTPs   5 μl
   - labelled triphosphate (0.5 mM) [a]   5 μl
   - 100 mM DTT   5 μl
   - DNase I, 1000 × diluted from a 1 mg/ml stock   5 μl
   - DNA polymerase I (5 U/μl)   5 μl
   - H₂O to a final volume of 50 μl   x μl

2. Mix and incubate at 15°C for 2 h [b]

3. Stop the reaction by adding 5 μl 0.5 M EDTA, pH 8.0.

4. Purify the labelled probe by Sephadex G-50 chromatography (column equilibrated with 10 mM Tris–HCl, 1 mM EDTA as the eluant). [c]

5. Add 1 μl glycogen or carrier DNA (herring sperm) to the eluate and precipitate with ethanol by adding 1/10 volume of 3 M sodium acetate, pH 5.6, and 2.5 volumes of cold ethanol.

6. Incubate at −20°C for 1 h and microcentrifuge at 15 000 g for 30 min at 4°C.

7. Dry the pellet and dissolve in either the appropiate buffer for the hybridizations or 10 mM Tris–HCl, pH 8.0, 1mM EDTA.

---

[a] Labelled triphosphates are mixed with unlabelled triphosphates in a ratio of 2:1 to 5:1 for efficient labelling. Final concentration should be 50 μM. By mixing differently labelled triphosphates in different molar ratios the multiplicity of the hybridization detection can be increased (42).

[b] Use an aliquot to check the probe size. For *in situ* hybridization the optimal fragment length is about 100–200 bases. If necessary add extra DNase to obtain the optimal fragment length.

[c] Flush the column with 50 μg herring sperm DNA or 0.1% SDS to avoid non-specific binding of labelled probe.

3.2.2 Random primed labelling

Random primed labelling (*Protocol 2*) is performed on a linearized, single-stranded DNA template. Oligonucleotides are randomly annealed, followed by enzymatic addition of labelled nucleotides. Probes generated by random labelling are composed of a larger portion of small fragments compared to nick-translated probes. Although greater incorporation of label can be achieved by random primed labelling the choice of method has to be determined empirically. In our experience, nick-translation of probes for fluorescent *in situ* hybridization is more convenient, since no linearization of the DNA prior to labelling is needed.

Protocol 2. Random primed labelling of DNA

Equipment and reagents

- DNA
- 10 × hexanucleotide mixture (10 × HM): 0.5 M Tris–HCl, 0.1M MgCl$_2$, 1 mM dithioerythritol, 2 mg/ml BSA, 62.5 A$_{260}$ U/ml hexanucleotides, pH 7.2
- 10 × deoxynucleotide triphosphate (10 × dNTPs): 1 mM dATP, dCTP, dGTP, dTTP with one triphosphate partially replaced by the labelled precursor[a]
- labelled nucleotides: e.g. digoxigenin-11-dUTP (DIG-dUTP), biotin-16-dUTP (bio-dUTP), fluorescein-12-dUTP (FITC-dUTP),

- biotin-7-dATP (bio-dATP) (see *Protocol 1* for suppliers)
- Klenow enzyme (2 U/μl)
- 0.5 M EDTA, pH 7.4.
- 3 M sodium acetate, pH 5.6
- Sephadex G-50, or spun columns
- carrier herring sperm DNA
- 1 × TE buffer: 10 mM Tris–HCl, pH 8.0, 1 mM EDTA
- ethanol (−20°C)

Method

1. Linearize and denature 10 ng–3 μg of DNA by heating for 10 min at 95°C then chill on ice for 5 min.

2. Mix the following in a microcentrifuge tube on ice:
   - 10 × HM (1 mM)   2 μl
   - 10 × dNTPs   2 μl
   - labelled triphosphates (0.5 mM)   2 μl
   - DNA up to   10 ng–3 μg
   - x μl sterile water up to a volume of 19 μl

3. Add 1 μl (2 U) Klenow enzyme.

4. Mix, microcentrifuge briefly and incubate for at least 60 min at 37°C.[b]

5. Stop the polymerase reaction by adding 2 μl 500 mM EDTA, pH 7.4.

6. Precipitate the labelled DNA by adding 1/10th volume of 3 M sodium acetate, pH 5.6, and 2.5 volumes of cold ethanol.

7. Keep at −20°C for 2 h and microcentrifuge at 15 000 g for 30 min at 4°C.

8. Wash the pellet with cold ethanol, carefully remove the solvent, and dissolve the pellet in, e.g., 1 × TE buffer. Store at −20°C until use.

[a] The ratio of labelled to unlabelled triphosphate is 1:2 for efficient labelling. Final concentration is 100 μM.

[b] Longer incubation increases the yield of labelled DNA. The size of the labelled DNA fragments is dependent on the length of the linearized DNA. If necessary reduce the fragment length, e.g. by sonication.

3.2.3 Polymerase chain reaction (PCR)

PCR technology has been introduced for *in situ* hybridization to obtain human specific probes from either rodent × human hybrid cell lines (29, 30), containing one human chromosome, or yeast artificial chromosomes (YACs) with human inserts up to 800 kb (31). In both cases human DNA can be specifically amplified *in vitro* using human specific primers to interspersed repetitive sequences (IRS-PCR; Alu- or Line-primers (*Protocol 3*) (32), (*Plate 1k*). Where probe DNA is limited, for example, flow-sorted chromosomes, microdissected chromosome bands, or even tumour DNA, sufficient probe material can be obtained by general DNA amplification using a degenerate oligonucleotide primer (DOP-PCR; ref. 33). The amplified DNA can be labelled by nick-translation or random priming, or by a second round of amplification, using similar PCR conditions, including a labelled nucleotide (29–33). In the latter case post-treatment of the labelled PCR-products with DNase I is necessary to obtain probes with the correct fragment length.

Protocol 3. Polymerase chain reaction (PCR)

Reagents

- DNA
- 10 × PCR buffer: 0.5 M KCl; 0.1 M Tris–HCl, pH 8.0; 15–25 mM MgCl$_2$; 0.1% gelatin [a]
- 100 × deoxynucleotide triphosphate (100 × dNTPs): 25 mM dATP, 25 mM dGTP, 25 mM dCTP, 25 mM dTTP [b]
- 100 × Alu [c] or DOP primer [d] (100–200 μM)
- *Thermus aquaticus* polymerase (*Taq*) (5 U/μl)
- paraffin or mineral oil

Method

1. Mix in a sterile screw-cap microcentrifuge tube in the following order:
   - H$_2$O to a final volume of   100 μl
   - 10 × PCR buffer   10 μl
   - 100 × dNTPs   1 μl
   - 100 × primer   1 μl
   - *Taq* polymerase   2.5 U

   mix well and add up to 2 μg DNA.

2. Cover the mixture with paraffin or mineral oil.

3. Carry out a number of amplification cycles in a thermal cycler as follows:
   - Denature initially for 5 min at 94°C and then:
     (a) *for Alu-PCR* 40 cycles of:
           denaturation for 1 min at 94°C,
           annealing for 1 min at 55–61°C, [e]
           extension for 4 min at 72°C;
     (b) *for DOP-PCR* 5 cycles [f] of:
           denaturation for 1 min at 94°C,
           annealing for 1.5 min at 30°C,
           extension for 4 min at 72°C. [g]

11

Protocol 3. *Continued*

- Follow by 35 cycles of:
    denaturation for 1 min at 94°C,
    annealing for 1 min at 62°C,
    extension for 3 min at 72°C, with an addition of 1 sec/cycle to
    the extension step.
- Finish with a final extension of 10 min at 72°C.

---

[a] The assay buffer conditions can be optimized by manipulating the Mg^{2+} concentration.

[b] If using PCR probes for labelling one of the triphosphates is replaced by the labelled precursor. See *Protocols 1* and *2* for details of labelled triphosphates.

[c] Several different Alu-primers have been described (31, 32).

[d] The DOP primer used is the 6 MW primer described by Telenius *et al.* (33).

[e] The annealing temperature depends on the Alu primer used (31, 32).

[f] For labelling purposes the five cycles at low annealing temperature are omitted in the second round of amplification (33).

[g] The original protocol (33) uses a 3 min transition from 30–72°C and 3 min at 72°C. However, several thermal cyclers do not have the option of temperature transition within a certain time-span. For this reason the conditions given here have been adjusted.

3.2.4 Primed *in situ* labelling (PRINS)

PRINS (*Protocol 4*) is a fast and simple procedure based on the incorporation of modified (biotin, digoxigenin, FITC) dNTPs into DNA synthesized *in situ* with oligonucleotides or denatured fragments of cloned DNA as primers (34–36). It allows the sequence-specific detection of repetitive DNA sequences, for example, centromeric, telomeric, and Alu-sequences (see *Plate 2l*). The detection of single copy sequences, however, remains problematic using the PRINS method (34–36).

Protocol 4. Primed *in situ* hybridization (PRINS) of DNA on metaphase spreads

Reagents

- standard spreads of metaphase chromosomes fixed in methanol:acetic acid (3:1) (see Chapter 2 this volume)
- 70% formamide, 2 × SSC (0.3 M sodium chloride, 30 mM sodium citrate), pH 7.0
- PRINS reaction mixture (25 μl/slide)[a] 1μM oligonucleotide[b], 10 mM Tris–HCl, pH 8.3, 50 mM KCl, 1.5–4 mM MgCl$_2$, 0.01% gela-

- tin, 50 μM of dATP, dCTP, and dGTP, 25 μM of Bio-11-dUTP, DIG-11-dUTP, or FITC-12-dUTP[c] (see *Protocol 1* for suppliers)
- *E. coli* DNA polymerase I, Klenow, or *Taq* polymerase
- 50 mM NaCl, 50 mM EDTA, pH 7.4
- graded alcohol series

Method

1. Dehydrate the slides in 70%, 90%, and 100% ethanol and air-dry.[d]

2. Denature the slides in 70% formamide, 2 × SSC, pH 7.0, for 2 min at 70°C.

3. Dehydrate in ice-cold 70% ethanol, followed by 90% and 100% ethanol, and air-dry.

4. Pre-incubate the slide, a 20 × 40 mm^2 coverslip, and the PRINS reaction mixture at the reaction temperature for 15–30 min.

5. Add 1 U *E. coli* DNA polymerase I, Klenow, or *Taq* polymerase to the reaction mixture, mix quickly, and spread the mixture under the cover-slip on the slide.[e]

6. Incubate in a moist chamber for 10–30 min at 45–55°C, or at 65°C in the case of *Taq* polymerase.[f]

7. Stop the reaction by immersing the slide in 50 mM NaCl, 50 mM EDTA at 65°C for 1 min.

8. Detect the incorporated and labelled nucleotides by means of immuno-cytochemical detection methods and standard fluorescence micro-scopy[g] (see Chapters 2–5 of this volume and *Table 3*).

---

[a] It is also possible to use 10 μl reaction mixture per slide by spreading the mixture with a 20 × 20 mm^2 coverslip and sealing with nail polish. After the PRINS reaction the nail polish can be removed with a scalpel after soaking the slides for 5–10 sec in chloroform.

[b] 1–10 μg per slide of a cloned DNA probe that has been linearized with an appropriate restriction endonuclease can be used as a primer in a PRINS reaction.

[c] In the case of a PRINS reaction with DIG-11-dUTP or FITC-12-dUTP the addition of 6.25 μM dTTP to the reaction mixture is recommended.

[d] Pre-treatment with 50–100 μg pepsin (2500–3500 units per mg protein) per ml 0.01 M HCl for 10 min at 37°C followed by post-fixation in 1% formaldehyde in PBS for 10 min at room temperature can be used to give better accessibility of the PRINS reagents to chromosomal DNA.

[e] The use of *E. coli* DNA polymerase I can result in more non-specific DNA labelling by comparison with Klenow, probably due to its additional 5′–3′ exonuclease activity.

[f] To improve the specificity of oligonucleotide annealing, the reaction mixture can be heated to 92–94°C for 5 min before adding the *Taq* polymerase (i.e. hot start). Then the mixture is spread with a coverslip on a pre-heated slide, after which the slide is incubated for 5 min at the appropriate annealing temperature and for 10–30 min at 72°C.

[g] Non-specific DNA labelling may be reduced by pre-incubation of the slides with T4 DNA ligase or Klenow in combination with dNTPs or ddNTPs.

3.2.5 *In vitro* transcription

Single-stranded RNA probes are provided by *in vitro* transcription, using plasmids which contain bacteriophage RNA polymerase-specific promoters (see *Protocol 5*, 37). Single-stranded probes have the advantage that they cannot reanneal in solution. RNA polymerase enzymes bind to promoter sites which have been inserted into the vector on both sides of the cloning site. They are extensively used for the detection of mRNA. The stability of RNA–RNA hybrids is greater as compared with RNA–DNA hybrids.

Table 3. Detection systems for non-radioactively labelled nucleic acid probes

Label	Immunochemical detection system/affinity detection system		
	Primary layer	Secondary layer	Tertiary layer
Biotin	avidin*		
	avidin*	biotinylated-anti-avidin	avidin*
	mouse-anti-biotin*	anti-mouse* IgG	
		biotinylated-anti-mouse IgG	avidin*
DIG	mouse-anti-DIG*	anti-mouse* IgG	
		biotinylated-anti-mouse IgG	avidin*
		DIG-labelled-anti-mouse IgG	anti-DIG*
	rabbit-anti-DIG*	anti-rabbit* IgG	
FITC	rabbit-anti-FITC	anti-rabbit* IgG	
	mouse-anti-FITC	anti-mouse* IgG	

DIG, Digoxigenin; FITC, fluorescein; IgG, immunoglobulin
* proteins can be conjugated with a fluorochrome or enzyme

Protocol 5. RNA labelling by *in vitro* transcription

Reagents

- linearized template DNA
- 5 × transcription buffer (5 × TB): 200 mM Tris–HCl, pH 7.5; 30 mM MgCl$_2$; 10 mM spermidine; 0.05% BSA; 50 mM DTT; 50 mM NaCl; 2 U/μl RNase inhibitor (human placental ribonuclease inhibitor)
- 10 × nucleotide triphosphate mixture (10 × dNTPs): 10 mM ATP, GTP, CTP, UTP) with one triphosphate replaced by the appropriate labelled precursor (see below)
- labelled triphosphates: e.g. digoxigenin-11-

UTP (DIG-UTP) (Boehringer–Mannheim), biotin-16-UTP (bio-UTP) (Sigma), fluorescein-12-UTP (FITC-UTP) (Boehringer–Mannheim); biotin-7-ATP (bio-ATP) (Amersham, UK)
- 0.5 M EDTA, pH 8.0
- 4 M LiCl
- cold ethanol (−20°C)
- SP6, T7, or T3 RNA polymerase
- TED buffer: 10 mM Tris–HCl, pH 8.0, 10 mM EDTA, 10 mM DTT

Methods

1. Mix the following in a microcentrifuge tube at room temperature:
   - 5 × TB — 4 μl
   - 10 × dNTPs — 2 μl
   - labelled triphosphate — 1 μl
   - linearized template DNA — 1 μg
   - sterile water up to a volume of 19 μl — x μl

2. Mix at room temperature and add 10 U of SP6, T7, or T3 RNA polymerase (usually 1 μl).

3. Incubate for 2 h at 37°C.

4. Stop the polymerase reaction by adding 2 μl 0.5 M EDTA, pH 8.0.

14

5. Precipitate the labelled RNA by the addition of 2.5 μl 4 M LiCl and 75 μl cold ethanol.

6. Keep at −20°C for 2 h and microcentrifuge for 15 000 *g* for 30 min at 4°C.

7. Wash the pellet with cold ethanol, carefully remove the supernatant and dissolve the pellet in TED buffer. Store at −20°C until use.

3.2.6 Oligonucleotide labelling

i. 3'-labelling

Several methods have been described for enzymatic labelling at the 3'-terminus of oligonucleotides. In general, they involve the enzymatic incorporation of modified nucleotides.

An enzymatic methodology analogous to the periodate method for chemical modification at the 3'-terminus of RNA has been developed (43). These authors have synthesized biotin-, fluorescein-, and tetramethylrhodamine-derivatives of *p'*-(6-aminohex-1-yl)-P2-(5'-adenosine) pyrophosphate, that could be used as substrates for T4 RNA ligase.

3'-O-(5'-phosphoryldeoxycytidyl) phosphorothioate and fluorescent 3'-O-(5'-phosphoryldeoxycytidyl)S-bimane phosphorothioate can be coupled to tRNA by RNA ligase (44). Oligo- and polynucleotides with a 3'-phosphorothioate group react readily with electrophiles as exemplified by the reaction with the fluorochrome monobromobimane.

A simple method has been described for the chemical labelling of DNA fragments at their 3'-terminus (45). The procedure involves enzymatic addition of 4-thiouridine by terminal deoxynucleotidyl transferase followed by a chemical modification of the incorporated nucleotide. Reactive haloacetamido derivatives are covalently attached by alkylation of the 4-thiouridine residue. In this way fluorescent dyes can be coupled to nucleic acids.

Protocol 6. Oligonucleotide 3' end labelling

Reagents

- 10 × terminal transferase buffer (10 × TTB): 1 M potassium cacodylate, 10 mM CoCl$_2$, 250 mM Tris–HCl, 2 mM DTT, pH 7.6, 2 mg/ml BSA
- 10 mM dATP
- labelled triphosphates (1 mM); FITC-dUTP, biotin-dUTP, digoxigenin-dUTP, or amino-hexyl-dATP

- oligonucleotide
- terminal transferase
- 4 M LiCl
- cold ethanol (−20°C)
- 0.2 M EDTA, pH 8.0
- TED buffer: 10 mM Tris–HCl, pH 8.0, 10 mM EDTA, 10 mM DTT

Method

1. Mix the following in a microcentrifuge tube:

- oligonucleotide   100 pM
- 10 × TTB   2 μl

Protocol 6. *Continued*

- dATP 1 μl
- labelled triphosphate [a] 1 μl
- terminal transferase (25–50 U) 1 μl
- distilled water up to a volume of 20 μl x μl

2. Incubate for 1 h at 37 °C.
3. Stop the polymerase reaction by adding 2 μl 0.2 M EDTA, pH 8.0.
4. Precipitate the labelled DNA by adding 2.5 μl 4 M LiCl and 75 μl ice-cold (−20 °C) ethanol.
5. Keep at − 20 °C for 2 h and microcentrifuge at 15 000 g for 30 min at 4 °C.
6. Wash the pellet with cold ethanol, carefully remove the supernatant and dissolve the pellet in TED buffer. Store at −20 °C until use.

[a] The ratio of labelled dUTP/unlabelled dATP is 1:10; the average tail length under these conditions is 50 nucleotides and 5 hapten molecules.

ii. 5′-labelling

Oligonucleotides can be synthesized containing an aliphatic amino group at the 5′-terminus (46). Protected amino-thymidine phosphoramidite must be used in the oligonucleotide synthesis and after deprotection, an aliphatic amino group — which is free for conjugation with different fluorochromes such as FITC and TRITC — is generated. Several amine-modified nucleotides have been prepared for appropriate chemical incorporation into DNA probes (47). Amine linker arms were chemically incorporated into oligonucleotides and the introduction of marker molecules such as fluorescein, dinitrobenzene, or biotin was achieved in this way.

The enzyme alkaline phosphatase can be coupled directly to short synthetic oligonucleotides (48). The oligonucleotides have a single modified base that contains a reactive amino group which enables covalent cross-linking, the enzyme being coupled using the homobifunctional reagent disuccinimidyl suberate. The resultant oligomer–enzyme conjugates hybridize to target DNA fixed to nitrocellulose and can be visualized by the alkaline phosphatase reaction. Although the melting temperature of the hybrids is reduced by up to 10 °C as a result of steric hindrance and mismatching, the probe remains specific and has a surprisingly high sensitivity.

4. Probes and detection systems

A wide range of probe modification procedures, both in the choice of the marker molecule (small molecules such as haptens or large molecules such as enzymes) and in the degree of nucleotide modification without notable effect on the hybridization efficiency, have been described. As is clear from the previous sections, nucleic acid polymers contain many different sites for the attachment of marker molecules. Modification of the nucleic acid probe in most cases changes the nature of the nucleotide base that is involved in base-pair formation.

4.1 Influence of probe labelling on hybridization efficiency and specificity

The formation of hybrids between a nucleic acid probe and DNA/RNA target sequences is a reversible process. The thermal stability of hybrids (melting temperature, T_m) is determined by a number of physical parameters such as salt and formamide concentrations as well as the length of the hybrid and their G/C content. Furthermore, non-radioactive labelling of nucleic acid probes often results in the formation of a base-pair mismatch at the site of the modified nucleotide. The latter also affects the stability of the hybrid between the modified probe and the DNA sequence. The stability of these hybrids has not been determined for all modifications. Nevertheless, some conclusions can be drawn regarding the type of base modification and its effect on the thermal stability of the final hybrid.

Small molecules — such as fluorochromes, haptens and biotin — affect the thermal stability only slightly when they are linked to terminal residues of the probe (1). For short probes (about 20 bases) the reduction in thermal stability will be more pronounced. Various authors report a reduction of the melting temperature of 1–5°C (13, 49).

Several chemical and enzymatic modifications, resulting in a random distribution of marker molecules throughout the nucleic acid probe, were shown to decrease the T_m by about 1°C per % modification. This indicates that about 5–10% of the nucleic acid bases can be modified without any significant effect on hybridization specificity, since 5–10% mismatch can normally be tolerated (2–4, 21).

Proteins such as enzymes can be coupled to nucleic acids over 1 kb in length with only a slight effect on hybridization efficiency (11, 12). In these cases large complexes were produced with a T_m value for hybrid formation up to 5°C lower than the T_m of the hybrids formed with non-protein labelled DNA probes. Apparently sufficient bases are still free for proper annealing.

When the enzyme alkaline phosphatase was covalently linked to short oligonucleotide probes (21–26 bases in length) (48), the T_m was decreased by approximately 10°C, as compared to non-modified probes. However, the specificity was still sufficiently high and a positive signal could be detected from only 50 fg of target DNA.

4.2 Influence of fragment length on hybridization efficiency

The choice of non-radioactive hybridization methods for the detection of target DNA or RNA sequences will depend on the system in which the method is to be used and the detection sensitivity that is required. For both non-radioactive filter hybridization and for *in situ* hybridization procedures the nucleic acid probes have to be modified to enable non-radioactive detection. However, in this respect there are some differences between the two methods. For *in situ* hybridization, the nucleic acid probes have to react with a nucleic acid target present within the protein matrix of the cell. In general, the presence of a protein meshwork which is often cross-linked results in reduction of hybridization efficiency, because of problems with the penetration of the probe and reduced presentation of target sequences. Additionally, in the interphase nucleus, as in chromosomes, most of the DNA is highly condensed. The DNA is not only associated with histones to form nucleosomes, but further condensation produces fibres of 100 and 300 Å, which are even more highly organized into a final, probably again helical configuration. Starting from the dsDNA strand, the degree of condensation is 5000- up to 250 000-fold. It is obvious that this striking difference from DNA immobilized on membrane calls for special pre-treatment steps when DNA probes have to be applied *in situ*. The probe size is also important for *in situ* hybridization. For non-radioactive *in situ* hybridization the optimal molecular weight has not always been reported and may depend on differences in the permeability of nucleic acid probe in the cell matrix as a result of various types of fixation (49). There are indications that a probe length of 50–100 bases is optimal. Sonication of the probes to a length of 200–600 is a practical compromise, leading to efficient *in situ* hybridization.

4.3 Detection systems

Detection systems for non-radioactively labelled nucleic acid probes are summarized in *Table 3*.

4.3.1 Filter hybridization

For non-radioactive hybridization procedures on solid supports one important goal, namely the replacement of the use of radioisotopes and autoradiography by immunochemical detection systems, has been achieved in the last decade (see *Table 4*). The sensitivity of non-radioactive protocols is now within the range of radioactive methods, and goes down to a detection level less than 0.1 pg. Procedures using enzymatic colour precipitation reactions (24), luminescent compounds (see Chapter 6 this volume), or time resolved fluorometry (50, 51) give a sensitivity in the range of 0.5–5 pg target sequence, and newer approaches include the ELISA systems using fluorogenic enzyme substrates (52), or affinity-based hybrid collection methods (50). Nowadays, synthetic probes containing one enzyme molecule (alkaline phosphatase) are used. With these probes, high sensitivity (50 fg target) is obtained when compared with radioactive labelled probes (48).

4.3.2 *In situ* hybridization

Great progress regarding the increase in sensitivity has also been made in the field of non-radioactive *in situ* hybridization. Target DNAs of about 5–15 kb

Table 4. Non-radioactive detection systems for filter hybridizations

Enzyme	Substrate	Product/detection
Peroxidase	H_2O_2/DAB	brown colour/visual absorption
	H_2O_2/chloronaphthol	purple colour/visual absorption
	H_2O_2/luminol/enhancer	chemiluminescent/X-ray film
Alkaline phosphatase	BCIP/NBT	blue colour/visual absorption
	naphthol-p/Fast Red	red colour/visual absorption
	naphthol-p/Fast Green	green colour/visual absorption
	AMPPD	chemiluminescent/X-ray film

DAB, diaminobenzidine; BCIP/NBT, bromo-chloro-indolyl phosphate/nitro blue tetrazolium; AMPPD[R], 3-(4-methoxyspiro{1,2-dioxetane-3,2′-tricyclo[3.3.1.1$^{3.7}$]-decan}4-yl)phenyl phosphate

can be detected routinely by this technique. This goal has been reached by effective labelling procedures, improved hybridization protocols, novel detection systems, and immunochemical amplification methods (see *Table 5*). Less progress has been made in the detection of low copy mRNAs or viral sequences as they are dispersed throughout the cellular compartment rather than spatially localized like chromosomal DNA. This implies that special approaches are necessary to detect and quantitate RNA. These adaptations require precautions to minimize the loss of RNA during fixation and hybridization, as well as optimization of hybridization conditions and of detection systems to obtain a higher signal to noise ratio (53). In the field of RNA *in situ* hybridization, procedures using radioactively labelled probes are more successful. However, the use of synthetic oligos for the non-radioactive detection of neuropeptide mRNA molecules in, for example, brain material have gained in impact and riboprobes are being increasingly employed (see Chapter 4 this volume). Probes can be routinely labelled with different labels and used for detection (38–40, see *Plate 1*).

4.3.3 Multiple-target detection systems

Simultaneous detection of several DNA/RNA targets on one filter or by *in situ* hybridization procedures in a single cell, chromosome spread, or tissue section is possible with a mixture of nucleic acid probes carrying different reporter molecules. These tags are then visualized with different, distinguishable affinity systems. The detection systems that can be combined for filter hybridization and *in situ* hybridization are summarized in *Tables 4* and 5.

Detection of multiple targets using non-radioactive filter hybridization is mostly based on the combination of peroxidase and alkaline phosphatase activity. In double-target hybridization methods these two enzymes are then detected with different substrates. Recently a triple-target hybridization system for filter hybridizations was developed based on one enzyme and three different substrates ('rainbow system', Boehringer–Mannheim). Between

Table 5. Non-radioactive detection systems for *in situ* hybridizations

Enzyme/label	Substrate/fluorochrome	Product or label/detection
Peroxidase	H_2O_2/DAB	brown colour/bright-field microscopy
	H_2O_2/chloronaphthol	purple colour/bright-field microscopy
	H_2O_2/TMB	green colour/bright-field microscopy
Alkaline	BCIP-NBT	blue colour/bright-field microscopy
phosphatase	naphthol-p/Fast Red	red colour/bright-field microscopy
		red fluorescence/fluo-microscopy
	naphthol-p/New Fuchsin	red colour/bright-field microscopy
		red fluorescence/fluo-microscopy
Fluorochromes	FITC	green yellow fluorescence/fluo-microscopy
	AMCA	blue fluorescence/fluo-microscopy
	TRITC; CY3	red fluorescence/fluo-microscopy
	Texas Red	red fluorescence/fluo-microscopy
	CY5	infrared fluorescence/fluo-microscopy

DAB, diaminobenzidine; TMB, tetramethylbenzidine; BCIP/NBT, bromo-chloro-indolylphosphate/nitro blue tetrazolium; FITC, fluorescein; AMCA, aminomethylcoumarin; TRITC, tetramethylrhodamine

each single DNA detection step the enzyme is inactivated and the next DNA target is then detected. By this approach a triple colour band staining in blue, green, and red is obtained. For *in situ* hybridization, double absorbant staining (ref. 54, see *Plate 1d*) can be employed. Alternatively, the number of different fluorescent dyes that can be used for detection is almost unlimited. Furthermore, the spatial resolution between two different ISH signals using different fluorochromes, for example fluorescein (FITC, yellow-green) in combination with rhodamine (TRITC or Texas Red, red) (see *Plate 1f,g,h*) or coumarin (AMCA, blue) is excellent (55, 56) (see *Plate 1i*). *Plate 1j* demonstrates how a translocation is detected in this way. In chronic myelogenous leukaemia (CML) the reciprocal translocation which produces the Philadelphia chromosome is characterized by fusion of the *bcr* gene on chromosome 22 and the *abl* oncogene on chromosome 9. By red labelling at the *bcr* locus and yellow-green labelling at the *abl* locus, the Philadelphia chromosome can be visualized in the interphase nucleus by co-localization of red and yellow-green fluorescence signals (57). This approach has already been used to detect, analyse, or characterize several translocations and marker chromosomes (see *Plate 1j*). For triple-target detection, the probes can be labelled with three different labels, for example, biotin, digoxigenin, and fluorescein, and detected with three different detection systems resulting in FITC, TRITC, and AMCA signals. For more than three different targets several different colours are required (58). This can be achieved by modification of a single probe with different labels simultaneously or by mixing differently labelled single probes, for one and the same target sequence, in the hybridization mixture. By this approach, (so-called 'ratio labelling'), a nucleic acid probe will stain the target in different colours (42). Probes can, nowadays, be labelled directly with different fluorochromes which simplify these analyses. Multiple-target *in situ* hybridization is used to demonstrate numerical as well as structural chromosome aberrations and to interrelate such complex genetic changes (see Chapter 2 this volume).

5. Conclusions

Although considerable progress has been made with respect to the non-radioactive *in situ* hybridization technique, the sensitivity of the methods is not yet sufficient to abolish radioactive *in situ* hybridization methods completely. For DNA targets the sensitivity needs to be increased by a factor of 10-fold in order to allow routine detection of unique (gene) sequences in chromosomes and interphase nuclei of about 1 kb or smaller. For RNA targets, precise data concerning the sensitivity of non-radioactive *in situ* hybridization in terms of the number of mRNA copies per cell are not available, but it is to be expected that relatively high mRNA levels need to be present for detection with *in situ* hybridization. For a non-radioactive hybridization procedure the sensitivity can be increased at several levels. The more marker molecules that can be coupled to hybrids formed between probe DNA and the nucleic acid target, the lower the detection limit will be. For this purpose a modification procedure is needed that allows a high labelling index of the probe without significantly influencing the hybridization efficiency. Furthermore, the formation of large immunological or nucleic acid networks, resulting in a high concentration of marker molecules at the target site could lower the detection limit. The combination of such sensitive cytochemical approaches with sensitive microscopic detection systems, such as, for example, laser scanning microscopy or microscopy using CCD cameras should also improve sensitivity.

Chemical and enzymatic approaches have advantages and disadvantages and the precise choice is determined by both facilities and experimental requirements.

References

1. Bauman, J. G. J., Wiegant, J., Borst, P., and Van Duijn, P. (1980). *Exp. Cell Res.*, **138**, 485.
2. Schroyer, K. R. and Nakane, P. K. (1983). *J. Cell Biol.*, **97**, 377.
3. Landegent, J. E., Jansen in de Wal, N., Baan, R. A., Hoeijmakers, J. H. J., and Van der Ploeg, M. (1984). *Exp. Cell Res.*, **153**, 61.
4. Tchen, P., Fuchs, R. P. P., Sage, E., and Leng, M. (1984). *Proc. Natl Acad. Sci. USA*, **81**, 3466.
5. Forster, A. C., McInnes, J. L., Skingle, D. C., and Symons, R. H. (1985). *Nucl. Acids Res.*, **13**, 745.

6. Hopman, A. H. N., Wiegant, J., and Van Duijn, P. (1986). *Nucl. Acids Res.*, **14**, 6471.
7. Verdlov, E. D., Monastyrskaya, G. S., and Guskova, L. I. (1974). *Biochim. Biophys. Acta*, **340**, 153.
8. Morimoto, H., Monden, T., Shimano, T., Higashiyama, M., Tomita, N., Murotani, M., Matsuura, N., Okuda, H., and Mori, T. (1987). *Lab. Invest.*, **57**, 737.
9. Draper, D. E. (1984). *Nucl. Acids Res.*, **12**, 989.
10. Renz, M. and Kurz, C. (1984). *Nucl. Acids Res.*, **12**, 3455.
11. Renz, M. (1983). *EMBO J.*, **2**, 817.
12. Syvanen, A. C., Alanen, M., and Soderlund, H. (1985). *Nucl. Acids Res.*, **13**, 2789.
13. Chollet, A. and Kawashima, E. M. (1985). *Nucl. Acids Res.*, **13**, 1529.
14. Chu, B. C. F. and Orgel, L. E. (1985). *DNA*, **4**, 327.
15. Agrawal, S., Christodoulow, C., and Gait, M. J. (1986). *Nucl. Acids Res.*, **14**, 6227.
16. Connolly, B. A. and Rider, P. (1985). *Nucl. Acids Res.*, **13**, 4485.
17. Hopman, A. H. N., Raap, A. K., Landegent, J. E., Wiegant, J., Boerman, R. H., and van der Ploeg, M. (1989). In *Molecular neuroanatomy* (ed. F. W. Van Leeuwen, R. M. Buijs, C. W. Pool, and O. Pach) p.43. Elsevier, Amsterdam.
18. Raap, A. K., Hopman, A. H. N., and Van der Ploeg, M. (1989). In *Techniques in immunocytochemistry* (ed. G. R. Bullock and P. Petrusz), Vol. 4, p. 167. Cambridge, Cambridge University Press.
19. Bauman, J. G. J., Pinkel, D., and Trask, P. J. (1989) In *Flow cytogenetics* (ed. J. Gray), p. 275. Academic Press, San Diego.
20. Matthews, J. A. and Kricka, L. J. (1988). *Analyt. Biochem.*, **169**, 1.
21. Langer, P. R., Waldrop, A. A., and Ward, D. A. (1981). *Proc. Natl Acad. Sci. USA*, **78**, 6633.
22. Dale, R. M. K., Livingston, D. C., and Ward, D. C. (1973). *Proc. Natl Acad. Sci. USA*, **70**, 2238.
23. Brigati, D. J., Myerson, D., Leary, J. J., Splholz, B., Travis, S., Fong, C. K. Y., Hsiung, G. D., and Ward, D. C. (1982). *Virology*, **126**, 32.
24. Leary, J. L., Brigati, D. J., and Ward, D. C. (1983). *Proc. Natl Acad. Sci. USA*, **80**, 4045.
25. Vincent, C., Tchen, P., Cohen-Solal, M., and Kourilsky, P. (1982). *Nucl. Acids Res.*, **10**, 6787.
26. Kessler, C., Holtke, H. J., Seibl, R., Burg, J., and Muhlegger, K. (1990). *Biol. Chem. Hoppe-Seyler*, **371**, 917.
27. Rigby, P. W. J., Dieckmann, M., Rhodes, C., and Berg, P. (1977). *J. Mol. Biol.*, **113**, 237.
28. Feinberg, P. and Vogelstein, B. (1984). *Anal. Biochem.*, **137**, 266.
29. Lengauer, C., Riethman, H., and Cremer, T. (1990). *Hum. Genet.*, **86**, 1.
30. Lichter, P., Ledbetter, S. A., Ledbetter, D. H., and Ward, D. C. (1990). *Proc. Natl Acad. Sci. USA*, **87**, 6634.
31. Breen, M., Arveiler, B., Murray, I., Gosden, J. R., and Porteous, D. J. (1992). *Genomics*, **13**, 726.
32. Nelson, D. L., Ledbetter, S. A., Corbo, L., Victoria, M. F., Ramirez-Solis, R., Webster, T. D., Ledbetter, D. H., and Caskey, C. T. (1989). *Proc. Natl Acad. Sci. USA*, **86**, 6686.

33. Telenius, H., Pelmear, A. H., Tunnacliffe, A., Carter, N. P., Behmel, A., Ferguson-Smith, M. A., Nordenskjold, M., Pfragner, R., and Ponder, B. A. J. (1992). *Genes. Chrom. Cancer*, **4**, 257.
34. Koch, J. E., Kolvraa, S., Petersen, K. B., Gregersen, N., and Bolund, L. (1989). *Chromosoma*, **98**, 259.
35. Koch, J., Mogensen, J., Pedersen, S., Fischer, H., Hindkjaer, J., Kolvraa, S., and Bolund, L. (1992). *Cytogenet. Cell Genet.*, **60**, 1.
36. Gosden, J., Hanratty, D., Starling, J., Fantes, J., Mitchell, A., and Porteous, D. (1991). *Cytogenet. Cell Genet.*, **57**, 100.
37. Kassavetis, G. A., Butler, E. T., Roulland, D., and Chamberlin, M. J. (1982) *J. Biol. Chem.*, **257**, 5779.
38. Larsson, L. I. and Hougaard, D. M. (1990). *Histochemistry*, **93**, 347.
39. Dirks, R. W., Van Gijlswijk, R. P. M., Vooijs, M. A., Smit, A. B., Bogerd, J., Van Minnen, J., Raap, A. K., and Van der Ploeg, M. (1991). *Exp. Cell. Res.*, **194**, 310.
40. Dirks, R. W., Van Gijlswijk, R. P. M., Tullis, R. H., Smit, A. B., Van Minnen, J., Van der Ploeg, M., and Raap, A. K. (1990). *J. Histochem. Cytochem.*, **38**, 4657.
41. Trainor, G. L. and Jensen, M. A. (1988). *Nucl. Acids Res.*, **16**, 11846.
42. Nederlof, P. M., Van der Flier, S., Vrolijk, J., Tanke, H. J., and Raap, A. K. (1992). *Cytometry*, **13**, 839.
43. Richardson, R. W. and Gumport, R. I. (1983). *Nucl. Acids Res.*, **11**, 6167.
44. Cosstick, R., McLaughlin, L. W., and Eckstein, F. (1984). *Nucl. Acids Res*, **12**, 1791.
45. Eshaghpour, H., Soll, D., and Crohers, D. M. (1979). *Nucl. Acids Res.*, **6**, 1485.
46. Smith, L. M., Fung, S., Hunkapillar, M. W., Hunkapillar, T. J., and Hood, L. E. (1985). *Nucl. Acids Res.*, **13**, 2399.
47. Ruth, J. L. (1984). *DNA*, **3**, 123
48. Jablonski, E., Moomaw, E. W., Tulles, R. H., and Ruth, J. L. (1986) *Nucl. Acids Res.*, **14**, 6115.
49. Kempe, T., Sundquist, W. I., Chow, F., and Hu ShiuLok (1985). *Nucl. Acids Res.*, **13**, 45.
50. Hemmila, I., Dakubu, S., Mukkala, V. M., Siitari, H., and Lovgren, T. (1984). *Anal. Biochem.*, **137**, 335.
51. Syvanen, A. C., Laaksonen, M., and Soderlund, H. (1986). *Nucl. Acids Res.*, **14**, 5037.
52. Nagata, Y., Yokota, H., Kosida, O., Takemura, K., and Kikuchi, T. (1985). *FEBS Letters*, **183**, 379.
53. Lawrence, J. B. and Singer, R. H. (1986). *Cell*, **45**, 407.
54. Hopman, A. H. N., Wiegant, J., Raap, A. K., Landegent, J. E., Van der Ploeg, M., and Van Duijn, P. (1986). *Histochemistry*, **85**, 1.
55. Nederlof, P. M., Robinson, D., Abuknesha, R., Wiegant, J., and Hopman, A. H. N., Tanke, H. J., and Raap, A. K. (1989). *Cytometry*, **10**, 20.
56. Nederlof, P. M., Van der Flier, S., Wiegant, J., Raap, A. K., Tanke, H. J., Ploem, J. S., and Van der Ploeg, M. (1990). *Cytometry*, **11**, 126.
57. Arnoldus, E. P. J., Wiegant, J., Noordermeer, I. A., Wessels, J. W., Beverstock, G. C., Grosveld, G. C., Van der Ploeg, M., and Raap, A. K. (1990). *Cytogenet. Cell Genet.*, **54**, 108.

58. Wiegant, J., Ried, T., Nederlof, P. M., Van der Ploeg, M., Tanke, H. J., Raap, A. K. R. (1991). *Nucl. Acids Res.*, **19**, 3237.
59. Kaa van de, C. A., Nelson, K. A. M., Ramaekers, F. C. S., Vooijs, G. P., and Hopman, A. H. N. (1991). *J. Pathol.*, **165**, 281.
60. Speel, E. J. M., Schutte, B., Wiegant, J., Ramaekers, F. C. S., and Hopman, A. H. N. (1992). *J. Histochem. Cytochem.*, **40**, 1299.
61. Hopman, A. H. N., Ramaekers, F. C. S., Raap, A. K., Beck, J. L. M., Devilee, P., Van der Ploeg, M., and Vooijs, G. P. (1988). *Histochemistry*, **89**, 307.
62. Poddighe, P. J., Moesker, O., Smeets, D., Awwad, B. H., Ramaekers, F. C. S., and Hopman, A. H. N. (1991). *Cancer Res.*, **51**, 1959.

2

Cytogenetic analysis

K. W. KLINGER

1. Introduction

In situ hybridization was first demonstrated in 1969 by Gall and Pardue (1). With further refinement (2), isotopic *in situ* hybridization became a useful tool for gene mapping. However, the relatively coarse map position generated by this method, coupled with the tedious and difficult nature of the technique, limited its widespread application. It is the advent of fluorescence *in situ* hybridization (FISH) that has revealed the true power of the technology, allowing fast, precise detection, and localization of single copy targets. FISH has rapidly achieved roles as a research tool, and as a diagnostic tool in the fields of prenatal diagnosis, cancer diagnosis, infectious disease diagnosis, and gene dosimetry.

Nucleic acid probe sequences can be hybridized *in situ* to their complementary DNA sequences within fixed chromosomes, extended DNA molecules, or cytoplasmic RNA or DNA (mitochondrial, bacterial, or viral). This chapter concentrates on metaphase and interphase cytogenetics, with a discussion of hybridization to extended DNA molecules as it relates to gene ordering.

FISH in metaphase cytogenetics involves the hybridization of a wide variety of probes (discussed in Section 4) generated by any of a number of methods to metaphase spreads produced by standard cytogenetic techniques. These chromosomes are relatively condensed, and each one is readily identifiable by size and banding pattern. In contrast, in the interphase nucleus the chromosomes are relatively extended, with no uniquely identifiable morphological features. However, each chromosome occupies a distinct focal domain in the interphase nucleus so that, after hybridization to a specific probe, a discrete hybridization signal is obtained in most nuclei for each specific chromosome present. For example, a cell exhibiting trisomy 21 yields three chromosome 21-derived hybridization signals in the nucleus, whereas a normal cell displays two. Thus, given a highly chromosome-specific probe, *in situ* hybridization can be used to determine how many copies of a given chromosome are present within the nucleus. In a similar manner, *in situ* hybridization can be used to determine the distance between two sites.

All nucleic acid hybridization assays are based on the principle of self-complementarity. The two strands of the DNA double helix are held together by hydrogen bonds between complementary bases. This complementary structure underlies DNA replication *in vivo*, and allows the use of one strand to detect the other specifically in a complex mixture of molecules *in vitro*. After the strands of the DNA helix are separated by heat or chemical denaturation, any two complementary molecules can 'reanneal', or hybridize. Thus, in all hybridization assays a labelled DNA or RNA probe is designed to be complementary to the target sequence of interest. When the probe and target are denatured, mixed, and allowed to reanneal, the hybridized probe allows detection of the target sequence. As with any hybridization assay, the fundamental specificity results from the specific sequence of the probe, and the stringency of the hybridization and wash conditions. The sensitivity depends on the target size (intensity of fluorescent signal is proportional to target), the precise fluor used for labelling, and the method of labelling and detection.

2. Principles of the assay

2.1 Theoretical considerations

There are four basic components to FISH analysis: (a) labelling of the probe, (b) preparation of the sample (target), (c) hybridization of the probe to the sample, and (d) detection of the bound probe. Each of these steps must be optimized for successful FISH analysis. Failure to do so will result in a decreased signal to noise ratio, and a concomitant loss of sensitivity.

In the first step, DNA probes complementary to the desired target are modified or 'labelled' with a reporter group. Probe labelling has been discussed in detail in Chapter 1 of this volume, and will be considered briefly in Section 5 of this chapter. The reporter group is later visualized using a detector labelled with one of a variety of fluorophores (3–8). Alternatively, fluorophores coupled directly to the probe molecules may be used. Fluorescent detection gives better spatial resolution than isotopic or enzymatic methods of detection, and allows multicolour analysis (Section 5).

Next, the target cells are processed and fixed, usually on to glass microscope slides. The goal of fixation is to preserve the target sequences in a form suitable for hybridization and generally involves acid fixation to remove basic proteins (*Protocols 2–5*). The labelled DNA probe and the target chromosomes are then denatured. If metaphase chromosomes are to be analysed, it is important that the denaturation of chromosomal DNA is accomplished without destruction of chromosome morphology. Following denaturation the probe is hybridized to the target sequence.

The hybridization solution usually contains carrier DNA and total genomic or Cot-1 human DNA to decrease non-specific binding of the probe in order to suppress non-specific hybridization. After the unbound probe is washed off, the fluorescent detector is added (unless directly labelled probes are used) and allowed to combine with the bound probe. The slides are washed to remove unbound reagents, and the reporter–detector complex is visualized by fluorescence microscopy (9–11) (*Protocol 7*). Non-isotopic techniques are now quite sensitive, but absolute efficiency, sensitivity, and specificity vary by laboratory protocol used, probe type, cell type, and degree of optimization.

2.2 Practical considerations: optimizing sample preparation

Specific protocols for metaphase and interphase sample preparation are given in Section 3. However, these protocols may need to be adjusted in order to achieve optimal efficiency for specific tissue or cell types, or to accommodate different probe designs. Limited efficiency may also result from constraints of probe composition, as well as variations in sample preparation and hybridization detection. This is particularly important in interphase cytogenetics where, in practice, inefficiencies mean that many nuclei may not hybridize and will, therefore, not be scoreable, and that a smaller than expected percentage of the countable nuclei will demonstrate the correct number of hybridization signals (10–13). Reproducible denaturation of DNA and penetration of probes/detection reagents (that is, efficient hybridization/detection) are very important factors in *in situ* hybridization. Over-denaturation or over-digestion with proteases causes loss of DNA from the nucleus, and distorts cellular morphology. On the other hand, cell types with extremely condensed nuclei (for example uncultured amniotic fluid cells, spermatozoa) require more aggressive treatment to allow efficient probe penetration.

These parameters can be determined empirically by carrying out the entire FISH protocol, and evaluating the quality of the ultimate result. However, this can be tedious if a large number of different parameters are to be tested. Alternatively, *in situ* nick-translation can be used to evaluate sample processing conditions, as described by Manuelidis (12) and in *Protocol 1*. This technique is particularly useful for evaluating DNA retention and accessibility to hybridization reagents.

Protocol 1. *In situ* nick-translation

Equipment and reagents

- nuclei or metaphase spreads
- 10 × nick-translation buffer (10 × NT): 0.5 mM dATP, dCTP, and dGTP (Pharmacia, LKB Technology), 100 mM β-mercaptoethanol, 500 mM Tris–HCl, pH 7.8, 50 mM Mg Cl$_2$ ^{a}
- biotin-11-dUTP (Enzo)
- DNase I (Amersham International)
- DNA polymerase I (Amersham International)
- H$_2$O
- 2 × SSC: 0.3 M sodium chloride 30 mM sodium citrate
- Avidin-FITC (Vector) (see *Protocol 7*)
- antifade ^{b}
- rubber cement

Protocol 1. *Continued*

Method

1. Prepare nuclei or metaphase spreads using the method under evalua-
tion (see *Protocols 2–6*).

2. On ice, make up the biotinylation reaction mixture by mixing the
following:
   - 10 × nick-translation buffer   10 µl
   - 1 mM biotin-11-dUTP (or desired derivatized nucleotide)   5 µl
   - 2 µg/ml DNAse 1 (freshly diluted in ice-cold water)   10 µl
   - 10 U/µl DNA polymerase 1   2 µl
   - dH$_2$O   73 µl

   Final volume =   100 µl

3. Spot 10 µl of reaction mixture on each hybridization site.

4. Add a coverslip, and seal with rubber cement.

5. Incubate at room temperature for 30 min.

6. Remove coverslips by submersion in 2 × SSC.

7. Carefully peel off the rubber cement. Dip the slide in 2 × SSC, float
off and discard the coverslip.

8. Wash and block the slides according to *Protocols 7C* and *7D*.

9. Detect according to *Protocol 7E*, using avidin–FITC as detection reagent.

10. If desired, nuclei can be counterstained using antifade containing
propidium iodide or DAPI.

[a] 10 × nick-translation buffer may be made in 10 ml quantities, split into 1 ml aliquots and stored at −20°C

[b] Antifade may be made from DABCO (1,4-diazabicyclo[-2.2-] octane) (Sigma). Dissolve 223 mg of DABCO in 800 µl of dH$_2$O, 200 µl Tris–HCl, pH 8.0, then add 10 µl of 100 µg/ml DAPI (4,6-diamidino-2-phenyl-indoyl) and 9 ml of glycerol. Wrap the tube in aluminium foil and mix overnight on a rocker platform. Divide into 1 ml numbered aliquots and store at 4°C in the dark. Use one numbered aliquot at a time. Propidium iodide may be used instead of DAPI.

Inefficient removal of protein (by protease treatment, acid fixation, etc.) may prevent adequate penetration of the probe resulting in failure of that sequence to hybridize to the tissue and consequent lack of signal. In the *in situ* nick-translation assay, inefficient probe penetration is indicated by a fluorescent 'ring' around the nucleus, while the centre remains dark. Overly harsh treatments (for example protease digestion or denaturation) can lead to DNA loss which appears as 'holes' against a background of overall fluor-escent staining of the nucleus. Ideally, after *in situ* nick-translation more than 90% of the nuclei should be completely, evenly, and brightly stained with the fluor. This result generally correlates with a high quality preparation for FISH analysis.

3. Sample preparation

3.1 Preparation of metaphase spreads

Metaphase spreads can be prepared by a wide variety of methods traditionally used for cytogenetic analysis. In general, chromosomes that are slightly less 'flat' than the ideal for GTG banding will give good hybridization results. Properly prepared metaphase spreads will show hybridization signals to both chromatids in the majority of metaphases. Two protocols for the preparation of metaphase spreads are described in this chapter, one for peripheral blood mononuclear cells (*Protocol 2*), and one for adherent cultured cells (for example, fibroblasts or amniocytes) (*Protocol 3*).

Protocol 2. Preparation of metaphase spreads from short-term blood cultures

Reagents

- Blood media: 100 ml 1640 RPMI, 20 ml fetal calf serum, 1 ml penicillin/streptomycin, 1 ml L-glutamine
- phytohaemagglutinin (PHA): add 5 ml sterile dH$_2$O into a vial of lyophilized PHA. Shake to mix.
- ethidium bromide (optional): stock 1 mg/ml
- in Hanks Balanced Salt Solution. Store at 4°C in the dark and handle with care.
- colcemid (stock 10 μg/ml)
- KCl (75 mM)
- fixative (3:1 methanol: glacial acetic acid) prepared immediately prior to use
- BUdR

A. Culture

1. Collect blood in an appropriate anticoagulant (e.g. sodium heparin, ACD, etc.).
2. Centrifuge the sample for 10 min at 290 g in a Centra 8 or equivalent centrifuge. Remove plasma layer and buffy coat with a sterile syringe, and dispense into a sterile tissue culture tube. Resuspend buffy coat in the plasma.
3. Add 5 ml complete blood media and 0.1 ml PHA to each tube.
4. Incubate for 72 h at 37°C.
5. If desired, add (0.1 ml) ethidium bromide solution to the tube 150 min prior to harvest. Mix and re-incubate at 37°C. If BUdR (final concentration 200 μg/ml) is to be included, add this 16–17 h before harvest).
6. Approximately 30 min before harvesting, add 50 μl of colcemid solution to each tube. Mix and place the tubes back in the 37°C incubator.
7. After 30 min, harvest the cultures by inverting the tubes to resuspend the cells. Transfer the cultures to labelled centrifuge tubes.

Protocol 2. *Continued*

8. Centrifuge the tubes for 10 min at 320 *g*.

9. Remove the medium by aspiration and resuspend the cells.

10. Add 1.0 ml of hypotonic solution (75 mM KCl) to each tube and mix gently.

11. Add 2 ml of hypotonic solution to each tube, again gently mix the tubes to resuspend the cells. Continue to add hypotonic solution to about 5 ml, mixing the tubes between each addition.

12. After 12 min at room temperature, centrifuge the tubes for 10 min at 160 *g* in a Centra 8 or equivalent centrifuge.

13. Remove the supernatant by aspiration and gently resuspend the cells. **NB** Failure to resuspend the cells completely at this point may result in clumping when the fixative is added.

14. Add 5–10 drops of freshly prepared cold fixative and bubble air gently through the cell suspension. Add the fixative one Pasteur pipette-full at a time in layers. Bubble gently after each pipette-full to resuspend cells. Continue to add the fixative in this manner until 5 ml have been added. Cap the suspensions and allow to sit for 15 min.

15. Centrifuge the cultures for 8 min at 180 *g*. Remove the fixative by aspiration. Fix a second time, centrifuge, aspirate, and add a third change of fixative for the preparation of slides or storage at 4°C. Pellets should be white at this point with no residual denatured haemoglobin (brown).

NB Always put the cells through at least three changes of fixative before storage.

B. *Metaphase spreads*

1. Prepare metaphase spreads after 3–5 changes of fixative, when the pellet is clean and white. If spreading is poor, refrigerate at 4°C overnight.

2. After harvesting, and when the pellet is clean and white, centrifuge for 8 min at 180 *g*. Aspirate the supernatant.

3. Add just enough fixative to make a slightly turbid cell suspension. The amount of fixative added will vary, therefore, upon the pellet size.

4. Clean the slides by soaking them in methanol in a large glass Coplin jar. Always have an excess number of slides in methanol. At the time of slide-making dip the slides into fresh methanol, agitate them, and drain them on paper towels until dry or dry with Kimwipes.

5. Drop 3 'drops' of the cell suspension on to a steamed slide (steam the slide by breathing on it or by holding the slide over a beaker of boiling

water) from an approximate height of 45–60 cm. Blow on the slide to help spread the chromosomes and then dry them in a humidity box or on the bench top. Check spreading by phase microscopy with a 10 × or 40 × objective: The metaphases should appear dark and evenly spread. If the metaphases appear underspread, correct by blowing harder, or dropping the suspension from a greater height. If the metaphases appear overspread, correct by not blowing so hard or by dropping the suspension from a lower point.

6. Allow slides to dry thoroughly for at least 4 h (or overnight if possible) on a hotplate at 50°C before proceeding with any staining technique.

Metaphase spreads from adherent cell cultures in flasks may be harvested by a trypsinization step followed by *Protocol 2*. Metaphase spreads from adherent cell cultures may be prepared according to *Protocol 3*.

Protocol 3. Preparation of metaphase spreads from adherent cultured cells

Reagents

- colcemid (10 mg/ml)
- KCl (75 mM)

- fixative (3:1 methanol:glacial acetic acid) prepared immediately prior to use

Method

1. Grow target cells in the appropriate media on sterile coverslips until colonies are forming and the cells are mitotically active. Mitotically active cultures have rounded-up, refractile cells.

2. Harvest the cells when a significant number of rounded-up cells are viewed in the colonies. Most cultures are harvested between 5 and 8 days.

3. Add 50 μl (10 mg/ml) colcemid to the culture and re-incubate for 25 min.

4. Remove the dishes from the incubator. Gently remove the medium with a Pasteur pipette from the edge of the dish.

5. Add 2 ml of hypotonic solution (75 mM KCl) and let stand for 20 min at room temperature.

6. Gently drop about 1–2 ml of fixative around the edge of the dish to the hypotonic solution. Allow to stand for 5 min.

7. Carefully aspirate the fixative/hypotonic mixture from the edge of the dish.

8. Add 2 ml of fresh fixative and allow to stand for 15 min. Aspirate the

Protocol 3. *Continued*

fixative and repeat this procedure with fresh fixative two or three more times.

9. Remove the last fixative by hand.

10. Place the lid of the dish underneath the dish, and place it in a controlled humidity chamber.[a]

11. When the cultures are completely dry, gently label the back of the dish using a fine point marking pen. Place the dishes in a 60°C oven overnight or at 90°C for 1 h.

[a] Humidity in the chamber should be approximately 45%. The fan blower should be set such that coverslips dry in about 90 sec. If no controlled humidity chamber is available, allow the coverslips to dry at room temperature. In this case, spreading may vary with environmental conditions.

3.2 Preparation of interphase nuclei

Interphase nuclei are present in most metaphase preparations, which can, therefore, be used for interphase analysis. However, one of the goals of interphase cytogenetics is the analysis of cells from uncultured or non-dividing preparations. Accordingly, two protocols are included for preparing this material. One is from short-term cultured lymphocytes (which can also be used for uncultured peripheral blood lymphocytes), while the other is for uncultured amniotic fluid cells (*Protocol 4*). These two cell types represent opposite ends of the spectrum of nuclear condensation. Cultured lymphocytes have relatively large, open nuclei, and the target DNA is readily accessible for hybridization. In contrast, the uncultured amniocyte nucleus is relatively small, very condensed, and does not allow good probe penetration. If processed according to the lymphocyte protocol detailed below, few nuclei hybridize, and many of the hybridized nuclei display an incorrect number of hybridization signals. Processed according to the optimized protocol (*Protocol 4*) greater than 90% of the nuclei hybridize, and 90–95% display the correct number of hybridization signals. The percentage of nuclei in a given sample which hybridize, and the extent to which hybridization reflects the correct genotype, is a complex product of probe design and performance, hybridization efficiency, and signal detection capability. It has previously been shown that the hybridization/detection efficiency of the assay has a disproportionate impact on the ability to detect the third signal in trisomic cells. If, for example, the aggregate ability to detect a target chromosome is 0.9, then the ability to detect two chromosomes in a nucleus is 0.9×0.9, or 0.81, and the ability to detect three chromosomes is $0.9 \times 0.9 \times 0.9$, or 0.729. It is obvious that high efficiencies must be achieved for *in situ* hybridization assays to have clinical utility in prenatal diagnosis, efficient multiprobe analysis, and for interphase gene mapping.

The following points are useful for the preparation of interphase nuclei from peripheral blood lymphocytes.

(a) Short-term lymphocyte cultures (see *Protocol 2*) can be used for interphase cytogenetics. Many interphase nuclei will be present.

(b) Check that the spreading conditions are appropriate for high quality interphase nuclei. If the nuclei are too 'flat' (so that poor probe penetration occurs) or too 'round' (i.e. yield an unacceptably three-dimensional nucleus with hybridization signals in different focal planes), adjust the dropping and 'spreading' conditions (see *Protocol 2*).

(c) When scoring the interphase signals after hybridization, remember that stimulated cultures will contain G_2 nuclei. These can be recognized by their larger nuclear volume, and the presence of paired signals for each probe.

(d) Interphase nuclei from uncultured blood lymphocytes can be prepared from buffy coats in the same manner, starting with step 8 of *Protocol 2A*. The chromatin will be more condensed compared to that obtained from stimulated short-term cultures.

Protocol 4. Preparation of interphase nuclei from uncultured amniotic fluid cells [a]

Reagents

• KCl (75 mM)
• fixative (3:1 methanol:glacial acetic acid) prepared immediately before use
• 30% fixative (3:1 methanol:glacial acetic

acid) in 75 mM KCl, freshly prepared in a glass container
• ethanol series (70%, 80%, 90%, 100%)

Method

1. Centrifuge the amniotic fluid at 550 *g* for 7 min at room temperature.

2. Aspirate the supernatant leaving approximately 50 μl of fluid above the pellet.

3. Resuspend the pellet to yield approx. 5×10^3 cells per slide.

4. Label the slides with the date and identifier, and mark two sample sites per slide by drawing circles on the back of each slide using a solvent resistant marker pen. Alternatively, masked slides may be used.

5. Place approx. 5×10^3 resuspended cells on the marked sample area.

6. Without disturbing the sample drops, place the slides flat in a humid box at 37°C for between 15 and 30 min, to allow the cells to settle.

7. Gently add KCl (pre-warmed to 37°C) to each sample site to yield a final concentration of 50 mM. Do not cause the sample to spread.

Protocol 4. *Continued*

8. Return the humid box to the 37°C incubator for an additional 15 min. Do not leave the samples at 37°C for more than 20 min.

9. Gently tip the slides sideways to allow the fluid to run off on to a paper towel. Do not allow the slides to dry. As each slide is drained, lay it flat on a clean paper towel and gently add 10 ml of 30% fixative in 75 mM KCl. Leave undisturbed at room temperature for 5 min.

10. Gently tip the fluid off the slides on to a paper towel. Immediately drop 5–6 drops of fresh undiluted 3:1 fixative on to each sample site from a height of about 60 cm.

11. Tip off any excess fluid and briefly blow on the slide.

12. Place the slides on a 60°C slide warmer for 5 min to dry.

13. When the slides are dry, dehydrate by passing through an ethanol series. Leave the slides in each ethanol concentration for 1 min at room temperature.

14. The slides are usually processed through hybridization immediately. However, if this is not possible or there are extra slides, store at −20°C until use.

[a] Interphase cells in G1 may be obtained from cell lines harvested at complete confluency.

Extended DNA molecules are useful for mapping studies by ISH (discussed further in Section 6.3). These can be produced by histone depletion and one suitable procedure is described in *Protocol 5*.

Protocol 5. Preparation of extended DNA molecules by histone depletion

Equipment and reagents

- 75 mM KCl containing 0.01% (v/v) Triton X-100
- chromosome isolation buffer (45): 1 M hexylene glycol, 0.5 M CaCl$_2$, 10 μM Pipes, pH 6.5 (46)
- histone depletion solution: 0.2 mg/ml dextran sulfate, 20 μg/ml heparin, 10 mM

  EDTA, 10 mM Tris–HCl, pH 9.0, 0.01% Nonidet P-40, 1 mM PMSF
- fixative (3:1 methanol:glacial acetic acid) prepared immediately before use
- 50% fixative (3:1 methanol:glacial acetic acid in dH$_2$O)

Method

1. Grow cells to confluence on glass coverslips in small Petri dishes.

2. Aspirate to remove the culture medium and add 75 mM KCl containing 0.01% (v/v) Triton X-100. Incubate for 20 min at 37°C.

3. Remove the KCl/Triton, then gently add cold chromosome isolation buffer from the edges. Incubate for 20 min at 4°C.

4. Remove the buffer, then, from the edges, gently add cold histone depletion solution. Incubate for 30 min at 4°C. **Note**: cells/DNA will tend to lift from the surface upon addition of the histone depletion solution. Handle the dishes extremely carefully throughout the remaining steps.

5. Carefully remove most of the solution (leaving the coverslip submerged), then gently add 50% 3:1 fixative in dH_2O. Allow to stand for 5 min.

6. Carefully remove the solution and allow the coverslip to dry.

7. Gently flood the coverslip with 3:1 fixative, and allow to stand for 10 min. Follow this with several changes of fresh 3:1 fixative.

8. Remove the fixative and allow the coverslip to air-dry.

4. Types of probe

Originally all probes used for FISH were derived from unique, single copy sequences, thereby avoiding gene sequences which may have several copies present along a chromosome, or chromosomes, or highly repetitive sequences. The technique of chromosomal *in situ* suppression hybridization (CISS) now allows the use of large genomic probes that contain interspersed repeat sequences (13–16).

Thus, a variety of types of molecular probes can now be used for FISH, including complex probes composed of the inserts from entire chromosome libraries (9, 15, 17–18), alpha satellite repeat probes (19), composite probe sets composed of single copy subclones (14), composite probe sets derived from inter Alu-PCR products, single cosmids (20), cosmid contigs, YACs (yeast artificial chromosomes) and BACs (bacterial artificial chromosomes). The choice of probe type varies with the specific analytic application.

4.1 Locus-specific probes

Locus-specific probes recognize a single locus, and generate bright but discrete signals. They can be generated in a variety of ways, and range from single copy probes such as cDNAs, PCR products, small inserts (plasmids or phage) to large insert genomic DNAs such as single cosmids, cosmid contigs, or YACs. The choice of single locus probe 'type' depends upon the application. Virtually all of the aforementioned probe types can be used for physical mapping on metaphase chromosomes using FISH, with the following restrictions. Small probes (less than 1–5 kb) may show faint signal on metaphase spreads, and insufficient signal in interphase nuclei, whilst probes over 5 kb generally give good signals. FISH analysis of cosmid probes will identify approximately 90% of the target sequences, with hybridization to both chromatids seen on 80% of the metaphase spreads. The use of smaller probes reduces the efficiency to 20–50% (*Plate 2a*). Interphase mapping of cosmid probes using FISH increases the resolving power, and is discussed in Section 6.2. Probes of cosmid size and above also have diagnostic applications. For example, in the case of a known characterized translocation, probes can be placed above and below the breakpoints. Labelled with different fluors, two paired signals are seen in the interphase nuclei if no translocation has occurred, whereas separated signals are seen after translocation. Such probes can also be used for chromosome enumeration in interphase nuclei. Since probes composed of cosmid contigs have a high efficiency of target detection without significant background 'noise' (8, 21–23), the signals are bright enough to be easily detected, and their very focal nature significantly improves trisomy detection compared to many other types of probes (Section 4.2) (*Plate 2b* and *c*). Finally, because these probes are locus specific, they can be designed to detect the presence or absence of any desired region of the chromosome. For example the region 21q22.3, the so-called 'critical region' for Down syndrome, and the Duchenne muscular dystrophy deletions.

4.2 Whole chromosome probes

Complex probes have been developed that consist of composite sequences derived from a single chromosome, and are used to 'paint' or decorate the entire length of the metaphase chromosome (15). Painting probes have been derived from the flow-sorted chromosome-specific libraries generated by the National Gene Library Project (14, 15, 17, 24, 25), and are available from the American Type Culture Collection (ATCC). Equivalent probes are now available from other sources. Subsets of chromosome paints can be generated by IRS-PCR (interspersed repeat sequences) from somatic cell hybrids (26), or by sequence independent amplification of microdissected chromosomal material (27, 28). Painting probes have their greatest clinical utility in the analysis of translocations and insertions (for example in cancer cytogenetics). They are also used in prenatal diagnosis to identify the origin of 'marker chromosomes' seen on cytogenetic (metaphase) analysis, but are less useful for prenatal detection of aneuploidy in interphase cells, since the complex probes generate a large, diffuse signal, which often has a low signal-to-noise ratio. This can make the edges of the signal difficult to delineate, and count. Overlap of the hybridization domains (domain coalescence) is also common, and results in the decreased ability to detect trisomy. Whole chromosome probes can also be used in investigations of nuclear topography, and as tools in gene mapping (see Section 7).

4.3 Repetitive probes

Probes for repeat sequences specific for only one, or at most a few, chromosomes can be extremely useful for mapping purposes (see Section 7), and may also have diagnostic use. Most of these repeat motifs are derived from the alpha satellite or satellite III families (19, 29–33). Chromosome-specific repeats label the target with very high efficiency, and generate a brilliant hybridization signal. Specific repeat sequences have been cloned for the human autosomes 1, 3–12, 15–20, and the sex chromosomes X and Y (34). Like whole chromosome probes, the chromosome-specific repeat probes are very useful for the identification of chromosomes in mapping studies and for the identification of marker chromosomes detected by cytogenetic analysis. These probes can also be used for aneuploidy detection in prenatal samples. However, the specificity of these probes are sensitive to hybridization conditions. If the conditions are not carefully controlled, cross-hybridization to other chromosomes may occur. The signal size is also sensitive to pericentromeric heteromorphisms. Perhaps the most serious limitation for prenatal diagnosis is the lack of a repeat sequence specific for chromosome 21 and chromosome 13. These two chromosomes, each of which causes significant disease when aneuploid, share the same centromeric repeat sequences, and cannot be differentiated using these probes. In this case the use of cosmid sequences specific for a particular locus on the chromosome is more useful (Section 4.1) (*Plate 2a*). Spatial resolution using specific repeat probes may be more problematic than locus specific probes in the interphase nucleus, but is substantially better than that of whole chromosome paints.

4.4 Comparative genomic hybridization (CGH)

CGH is a relatively new procedure that allows entire genomes to be screened for changes in copy number. Genomic DNA from the 'test' individual is labelled and detected with a fluorophore, standard genomic DNA is labelled with a second fluorophore, and both DNAs are hybridized to reference metaphase spreads. The ratio of fluor 1:fluor 2 is indicative of no loss or gain (~1:1), loss (<1:1), or gain (>1:1) of genetic material. First described by Kallioniemi *et al.* (35) the technique is seeing increasing use as a tool to search for probable sites of tumour suppressor genes or oncogenes, for detection of aneuploidies, and for characterizing unbalanced genetic material.

5. Probe labelling

Probe labelling has been discussed in detail in Chapter 1 of this volume. In this section a general scheme for probe labelling is presented, with the rest of the discussion concentrated on multicolour fluorescence assays. It is important that sufficient label is incorporated in order to yield a high specific activity, and also that the mean size of the finished probe is in the order of 200 base pairs. Larger probe fragments contribute significantly to background in FISH assays. In *Protocol 6* a radioactive tracer reaction is included to monitor incorporation of the desired label, and final probe size. The final probe size can also be estimated from running the DNA alongside known markers on a 2% ethidium bromide stained agarose gel. Although *Protocol 6* describes the incorporation of a biotinylated nucleotide, any derivatized nucleotide can be incorporated using this procedure. However, the polymerase has a lower tolerance for some derivatized nucleotides, which must be diluted with non-derivatized nucleotide for efficient incorporation. The optimal concentration of derivatized nucleotide can be determined by running a test series of various ratios of derivatized to non-derivatized nucleotide (for example 1:1, 1:2, 1:3, 1:4).

Protocol 6. Generalized probe labelling

Equipment and reagents

- [α-^{32}P]dATP
- DNase I (Sigma, Amersham International)
- glycerol
- 10 × nick-translation buffer (see *Protocol 1*)
- biotin-11-dUTP (Enzo)
- DNA polymerase I (New England Biolabs, Amersham International)
- 500 mM EDTA, pH 8.0
- Sephadex G-50-80
- DE81 filters
- 2-ply GFC filters
- 0.5 M NaHPO$_4$
- 1 × TE
- SDS
- Spun column buffer: 1 × TE, 0.2% SDS. Sterilize through a 0.2 µM filter and store at room temperature

A. *Biotinylation reaction*

1. Thaw out the [α-^{32}P]dATP behind a shield in the 'hot' lab. before starting the reaction. This will take about 30 min.

2. Dilute 5 µl of the 1 mg/ml stock of DNase I in 50% glycerol into 245 µl of ice-cold distilled water to give a 20 µg/ml stock solution. For each biotinylation reaction, dilute 15 µl of this 20 µg/ml stock solution into 135 µl of ice-cold distilled water to give a working solution of 2 µg/ml DNase I (prepare one master tube for all reactions and aliquot from this). These dilutions should be done immediately before using the DNase I.

3. Label one 15 ml conical tube (for the biotinylation reaction) and one 1.5 ml microcentrifuge tube (for the ^{32}P tracer reaction) for each probe DNA to be biotinylated, and place both tubes on ice and allow to cool.

4. Prepare multiples of the following reaction mix in the 15 ml conical tube on ice. Up to a 6-fold reaction mixture for a single probe set can be done at one time.
   - DNA .... 20 µg
   - 10 × nick translation/biotinylation buffer ... 100 µl
   - 600 mM biotin-11-dUTP .... 50 µl
   - 2 µg/ml DNAse I .... 100 µl
   - 10 U/ml DNA polymerase I .... 20 µl
   - distilled water .... to 1000 µl

5. Remove a 10 μl aliquot of the biotinylation reaction and put into the 1.5 ml microcentrifuge tube for the ^{32}P tracer reaction. See Part C for the details of this reaction.

6. Incubate the remainder of the biotinylation reaction at 16°C for 2 h. (Incubation time range: 1 h 45 min to 2 h 15 min.)

7. At the end of the incubation return the reaction tube to ice and add 100 μl of 500 mM EDTA, pH 8.0, per 1 ml reaction.

8. Incubate the reaction in a waterbath at 65°C for 15–20 min.

9. Briefly spin the reaction tube to get all the solution to the bottom.

10. Apply the 1 ml of the biotinylation reaction to the centre of a spun column of Sephadex in a 5 ml syringe. See Part B for preparing the spun column. Use one column per 1 ml of reaction mix.

11. Centrifuge the column for 3 min at 320 *g*.

12. Transfer the eluate from each column into a fresh 15 ml conical tube.

13. Measure the optical density at 260 nm (OD_{260}) of the undiluted DNA from the spun-column using the spun-column buffer as a blank.

14. Use the following formula to calculate the concentration of DNA in the sample. An OD_{260} of 1 corresponds to 50 μg/ml of DNA.
    => OD_{260} × 50 = μg/ml DNA (equivalent to ng/μl).

15. Transfer 1 ml aliquots of the DNA into 1.5 ml microcentrifuge tubes. Write the probe name, batch number, and concentration on the tube and store at −20°C.

B. *Preparation of spun-columns*

During the nick translation/biotinylation incubation prepare one 5 ml spun-column in a 5 ml syringe for each biotinylation reaction.

1. Remove the plunger from a syringe and place a 2-ply GFC filter into the bottom of the barrel.

2. Pipette G 50–80 Sephadex in 1 × TE into the barrel and allow the resin to begin settling. Continue adding Sephadex until the packed resin is at the top of the syringe – the best column is formed when the resin settles in a continuous bed rather than in layers, so do not let the column run dry until the syringe is full.

3. Remove the cap from a microcentrifuge tube and put it in a 15 ml conical polypropylene tube. Place the packed syringe in the 15 ml tube with the outlet inside the microcentrifuge tube. Spin for 3 min at 320 *g*. Discard the supernatant.

4. Wash the column with 1 ml of 0.2% SDS/1 × TE, spin in the clinical centrifuge for 3 min at 320 *g* and discard the supernatant. Repeat this wash step once more.

Protocol 6. *Continued*

5. Place the syringe in a fresh 15 ml conical polypropylene tube contain-ing a clean microcentrifuge tube without cap, before adding the nick translation/biotinylation reaction.

C. *^{32}P tracer reaction*

1. To the 10 µl of the biotinylation reaction removed in Step A5, add 0.5 µl of 10 mCi/ml [α-^{32}P]dATP (5 µCi) and mix.

2. Incubate the reaction at 16°C alongside the biotinylation reaction for the same length of time.

3. At the end of the incubation, return the reaction tube to ice, pulse-spin it in the 'hot' lab microcentrifuge to get any condensation down to the bottom of the tube, and return the tube to ice.

4. Spot 5 µl of the reaction on to each of two DE81 filters labelled T0 and T1 and allow to air dry.

5. Wash the T1 filter 3 times in 0.5 M NaHPO$_4$ for 5 min per wash. All washes use approximately 20 ml of the appropriate solution in a 100 ml beaker. These washes remove unincorporated label from the filter, leaving the labelled double-stranded DNA fragments bound to the filter, and gives the total number of ^{32}P counts incorporated into the DNA.

6. Dry both the washed filter (T1) and the unwashed filter (T0) in the vacuum oven for 5 min.

7. Place the filters in separate plastic scintillation vials, add sufficient scintillation fluid to cover the filter, and cap the vial.

8. Count both filters in a scintillation counter.

9. Calculate the % incorporation of ^{32}P label into the DNA using the following formula:

$$\frac{\text{c.p.m. for filter T1}}{\text{c.p.m. for filter T0}} \times 100 = \% \text{ incorporation.}$$

5.1 Multicolour FISH

Three methods of multicolour labelling are commonly used for FISH analysis, (a) single labelling of each probe (currently up to three colours), (b) combina-torial, and (c) ratio labelling for simultaneous analysis of a larger number of probes.

A variety of directly conjugated fluorescent nucleotides are now available, including Fluorescein isothiocyanate-(FITC), Texas Red-(TR), rhodamine-, and coumarin-(AMCA) conjugates. Probes can be labelled with these nucleo-tides, hybridized, and washed appropriately, and the hybridized signal visualized either by epifluorescence microscopy using appropriate filter sets, or by laser microscopy using lasers matched to the fluors. Alternatively, an indirect labelling method can be used in which the probes are labelled with haptens, such as biotin, dinitrophenol (DNP), or digoxigenin (DIG), and detected by indirect methods (see Chapter 1 this volume). The direct labelling method gives lower background than indirect labelling, but also yields less signal intensity. If the slides are to be read manually the absolute signal intensity is of paramount importance. However, with digitized imaging systems the chief parameter is signal to noise ratio, and direct labelling may be the appropriate choice.

Combinatorial labelling increases the effective number of detector 'fluors' available, and, therefore, the number of targets that can be analysed compared to the number of individual fluorophores (36, 37). In this approach each probe is labelled with 1, 2, or 3 fluors at roughly equimolar ratios. For example, seven probes could be resolved with three hapten/fluor pairs as follows: probes 1, 2, and 3 would be labelled with pure fluorophore, whilst probe 4 would be labelled with fluor A and B, probe 5 with A and C, probe 6 with B and C, and probe 7 with A, B, and C. Similarly, four hapten/fluor pairs would allow the detection of 15 targets. Combinatorial labelling schemes rely on computer-supported digital imaging camera systems to collect, align, pseudocolour and display the images. Ratio labelling differs from combinatorial labelling in that, after the probes are labelled, they are mixed in defined, multiple ratios, rather than the approximate equimolar ratio of combinatorial labelling (38). If fluors such as AMCA, TRITC, and FITC are used in ratios varying from pure fluor to 5:1, a range of colours are generated which can be resolved by the human eye. Thus digital imaging and pseudocolouring can be avoided. However, if infrared fluors are used, image-processing support is again required. Ratio labelling works well for whole chromosome paints, but is not suitable for cosmid or YAC probes. Since the amount of single copy DNA varies from clone to clone, and is unknown in most cases, it would not be possible to construct the defined ratios required by this technique.

Protocol 7. *In situ* hybridization (ISH)

Equipment and reagents

- labelled probe (see *Protocol 6*)
- human Cot-I DNA (BRL)
- sonicated salmon sperm
- 3 M sodium acetate, pH 5.3
- ethanol
- formamide
- 20% dextran sulfate in 12 × SSC (1.8 M sodium chloride, 0.18 M sodium citrate) pH 7.0
- Speed-Vac

A. *Probe cocktail preparation*

1. Count the number of sites to be hybridized with each probe (*n*) and prepare sufficient probe for *n* + 2 sites. A typical cocktail will contain

---

Protocol 7. *Continued*

> 2 μg human Cot-1 DNA and 8 μg salmon sperm DNA (sonicated to ~ 200 bp in size) and 10–100 ng of labelled probe.

2. For each probe cocktail aliquot the appropriate volumes of probe, Cot-1 DNA, and salmon sperm DNA into a tube. Add 1/10th of the final volume of 3 M sodium acetate, pH 5.3. Add 2 volumes of −20°C absolute ethanol and mix.

3. Pellet the DNA in the microcentrifuge at 4°C for 10 min at 7000 *g* and carefully pour off the supernatant.

4. There must be a visible pellet after this microcentrifuge spin, even if sufficient probe for only one site is being precipitated. If no pellet is visible, discard and start again at Step 1.

5. Add 1 ml of cold 70% ethanol to the pellet and vortex briefly. Spin in the centrifuge for 5 min to re-pellet the DNA and then gently pour off the supernatant.

6. Dry the pellet in a Speed-Vac for 5–15 min. Check the pellet periodically as it should not be allowed to over-dehydrate; otherwise it will be difficult to resuspend.

7. There must be a visible pellet after drying in the Speed-Vac, even if sufficient probe for only one site is being precipitated. If no pellet is visible, discard and start again at Step 1.

8. Resuspend the pellet in half the final volume (5 μl per sample spot) of 100% deionized formamide and leave at room temperature for at least 4 h followed by heating the DNA in the formamide for 5 min at 65°C. It is essential that the DNA is completely resuspended in formamide before proceeding to the next step.

9. Add half the final volume (5 μl per sample spot) of pre-warmed 20% dextran sulfate in 12 × SSC and vortex vigorously to mix. **Note** if dextran sulfate is added before the DNA is completely resuspended in the formamide the DNA will never go into solution.

10. Spin the probe cocktail briefly in a microcentrifuge to get all the solution to the bottom of the tube.

B. *Hybridization procedure*

Equipment and reagents

- ethanol series (70–100%)
- rubber cement/glue or cow gum
- slide warmer

1. Normally slides will be hybridized immediately following processing and the hybridization procedure will start at Step B4. However, slides which have been stored at −20°C must be warmed to room temperature and re-dehydrated before use.

2. Slides which have been stored at −20°C can be left on the bench to warm up or placed on the 60°C slide warmer until dry. If a slide warmer is used do not leave the slide on it any longer than is necessary to dry the slide.

3. Dehydrate the slides through the ethanol series by incubating them for 1 min in each of the 70%, 80%, 90%, and 100% ethanol solutions.

4. After the last dehydration step, allow the slides to air-dry either at room temperature or on a slide warmer at 60°C. **Note**: From Step 5 through the rest of the hybridization and post-hybridization procedures the slides should never be allowed to dry.

5. Pipette 10 µl of the appropriate probe cocktail in hybridization solution on to one sample spot on the slide.

6. Carefully lower the coverslip on to the hybridization solution and gently squeeze out the bubbles.

7. Seal the coverslip to the slide using rubber cement.

8. Place the slide on the 80°C slide warmer for 7 min (absolutely *no longer* than 9 min) to denature the sample and probe DNA.

9. Incubate the slide in a moist chamber overnight (16–20 h).

C. *Washing the slide*

Equipment and reagents

- 2 × SSC
- 50% formamide/2 × SSC, pH 7.0 (mix equal volumes of formamide and 4 × SSC (0.6 M sodium chloride, 60 mM sodium citrate), adjust pH to 7.0)
- 0.1 × SSC
- Coplin jars

Note: Once the washing procedure is started the slide must not be allowed to dry out.

1. Dip the slide in 2 × SSC and carefully peel the rubber cement off the slide and coverslip. Dip the slide in 2 × SSC again and carefully slide off the coverslip and discard.

2. Place the slide in the first of the Coplin jars containing the 50% formamide/2 × SSC. Repeat the process of removing the coverslips from four more slides. Process five slides per Coplin jar and no more than ten slides at a time.

3. Wash the slides in each of three pre-warmed Coplin jars containing 50% formamide/2 × SSC for 5 min (range 4–6 min) at 42°C in a waterbath. (Temperature is critical.) *This step should be performed in a fume hood.*

4. Wash the slide in a Coplin jar containing 2 × SSC for 2 min (range 15 min) at room temperature.

Protocol 7. *Continued*

5. Wash the slide in each of three pre-warmed Coplin jars containing 0.1 × SSC for 5 min (range 4–6 min) at 60°C in the waterbath.

6. After the last wash, place the slides in 2 × SSC at room temperature.

7. For direct detection of signal proceed to step F1. If an indirect detection method is used, the slides must first be blocked to prevent non-specific binding — proceed to step D1.

D. *Blocking the slides*

Reagents

- 2 × SSC
- 4 × SSC (see *Part C*)
- 3% BSA (bovine serum albumin)/ 4 × SSC

1. Remove one slide at a time from the 2 × SSC (Step C6) and remove excess buffer by gently flicking the slide.

2. Add 100 μl of 3% BSA/4 × SSC per sample spot and carefully place a 22 × 30 mm coverslip over the solution taking care not to allow air bubbles to form under the coverslip.

3. Repeat Steps 1 and 2 for the remaining slides in the series, treating only one slide at a time.

4. Once all the slides have been treated, incubate them with the blocking solution at 37°C in a pre-warmed moist chamber for at least 30 min and no longer than 60 min.

E. *Detection of signal, indirect (biotinylated probe)*[a]

Reagents

- Avidin–FITC (Vector)
- 4 × SSC, 1% BSA, 0.1% Tween-20
- 4 × SSC (see *Part C*), 0.1% Tween-20
- 2 × SSC (see *Part C*)

1. After incubation in blocking solution (Step D4), carefully remove the coverslips from one slide and remove the excess solution by gently flicking the slide. There is no need to remove all the excess solution.

2. Prepare a working solution of 5 μg/ml avidin–FITC in 4 × SSC, 0.1% BSA, 0.1% Tween-20. Spin the diluted solution before use for 5 min in a microcentrifuge at room temperature. [b]

3. Add 100 μl of working concentration avidin–FITC to each sample spot and gently place a 22 × 30 mm coverslip over the sample, again taking care not to leave air bubbles under the coverslip.

4. Repeat Steps 1 and 3 for the rest of the slides in the batch, handling only one slide at a time.

5. Once all the slides in the batch have been treated, incubate them at 37°C for 30 min in a pre-warmed moist chamber covered with aluminium foil. (Incubation should be for at least 20 min, but no longer than 40 min.)

6. Remove the slides from the humid box one at a time, gently remove the coverslips, and place the slide in the first pre-warmed Coplin jar containing 4 × SSC/0.1% Tween-20.

7. Once all the slides are in the first of the pre-warmed Coplin jars containing 4 × SSC/0.1% Tween-20 at 42°C in the waterbath, wash for 5 min (range 4–6 min). Repeat this wash in each of the subsequent two Coplin jars containing 4 × SSC/0.1% Tween^r-20 for 5 min each wash.

8. Place the slides in 2 × SSC at room temperature and leave for at least 2 min. (Upper time range is not critical, however the slides should be exposed to light as little as possible.)

9. Proceed to Step F1 to mount slides.

[a] Once the detection procedure has been started the slides should be exposed to light, especially fluorescent light, as little as possible.

[b] If the fluorochrome conjugated antibody is kept at room temperature for too long, fluorochromes can be released resulting in an increase in background. Free fluorochromes can be removed from solution by purification over a spin-column.

F. *Mounting*

Reagents

- 2 × SSC (see *Part C*)
- Antifade (see *Protocol 1*)
- nail varnish

1. Remove one slide at a time from the 2 × SSC (Step E8) and remove the excess buffer by gently shaking the slide then blotting the edge on to a paper towel.

2. Add 10 μl of Antifade/DAPI to each sample spot and carefully place a 22 × 22 mm coverslip on top.

3. Carefully press the coverslip on to the slide to expel any air bubbles and aspirate expelled Antifade from around the coverslip.

4. Place the slide with the coverslip down on a paper towel and gently press the slide with a second paper towel to absorb any excess Antifade.

5. Seal the coverslip on to the slide with nail varnish.

6. When the nail varnish is dry, remove remaining excess Antifade by dipping the slide briefly into a Coplin jar containing distilled water and dry by gently pressing the slide between two paper towels.

7. Store the slides at −20°C in the dark until they are read, and store the same way immediately after reading.

6. Gene mapping

FISH can be used a a tool for physical gene mapping at three levels of resolution: metaphase chromosomes (~1–2 Mb), interphase chromosomes (40 kb–1 Mb), and extended DNA molecules (5 kb–100 kb). A combination of suppression hybridization and multicolour analysis allows rapid ordering of genomic clones.

6.1 Metaphase mapping

Probes can be ordered along the length of the chromosome in terms of relative order, band position, or fractional length of the chromosome. Relative order among a group of probes can readily be determined by simultaneous hybridization and multicolour analysis. Alternatively, clones can also be ordered in groups of three using two-colour analysis: two clones are labelled with one fluor, while the third is labelled with the alternate fluor. Hybridization performed in the various pairwise colour combinations allows unambiguous assignment of probe order.

Chromosomal identification can be achieved by simultaneous hybridization to a fiduciary marker or a whole chromosome paint, by simultaneous banding, or by banding post-hybridization. Chromomycin, quinacrine, Hoechst 33258, DAPI, and propidium iodide staining can be combined with the hybridization protocol to allow assigment to a specific band. Metaphase spreads can be stained with Giemsa after FISH analysis, to generate a classical GTG banding pattern. However, the banding pattern is often of poor quality, and always requires carrying out the FISH and banding procedures in two steps. This is time consuming, and relocation of the chromosomes is tedious.

Map position may also be determined by measuring the distance from the tip of the chromosome to the hybridization signal relative to the entire length of the chromosome or arm (fractional length). Subregional map assignments made by fractional length measurements allow higher resolution ordering, since order may be established for multiple probes that map within the same cytogenetic band. Chromosome condensation will influence the resolution achieved by this technique: the more elongated the chromosome the greater the resolution. Intrachromosomal variations will also affect resolution. Since probes are ordered along the length of the chromosome, signals which resolve across the width of the chromosome are not useful for ordering. The precise limits of resolution may also vary with the position on the chromosome, as the DNA may be more tightly packed in some subregions than others.

6.2 Interphase mapping

Hybridization to the relatively decondensed chromatin of the interphase nucleus allows the order and distance to be measured between probes mapping between 25 and 250 kb apart, and perhaps as far as 1 Mb apart (11, 39). Distances within this range are relatively linear, whereas over longer distances the looping back on itself of the chromatin fibre disrupts the linear relationship. Probes which are known to map to the same chromosomal subregion can be readily ordered by interphase mapping. As with metaphase mapping, probes are labelled with at least two colours, and the number of nuclei with (for example) a red, red, green orientation versus red, green, red are scored. Obviously a large number of nuclei must be counted in order to achieve a significant answer. Again, either repeating the analysis with the various pairwise combinations or use of multicolour analysis improves the accuracy. The distances between the signals can be measured on screen if computer-supported image analysis is used. Alternatively, photographic images can be projected on a wall or screen, and the distances measured with a ruler. Resolution may be somewhat improved by mapping in hybrid pronuclei (40, 41). However, this is an extremely difficult technique, and only improves resolution by approximately 20 kb.

6.3 Mapping to extended DNA molecules

Further resolution can be achieved by hybridizing to DNA 'extended' from its interphase state. These methods include free chromatin mapping (42), liberation of DNA from the nucleus in halo-like loops (halo mapping, 43), and generation of extended DNA molecules by histone depletion (44). The procedure described in *Protocol 5* yields DNA molecules of varying degrees of decondensation, with the maximum extension seen equivalent to that of BDNA. Overlaps and gaps of 1.5 kb can readily be detected using this technique. The upper end of the range is still undefined, but is likely to be of the order of several hundred kilobases (*Plate 2e*). Tightly linked probes (for example members of a YAC or cosmid contig) can easily be ordered. Again, at least two-colour fluorescence is required, and multicolour analysis is helpful.

7. Image analysis

Routine FISH analysis can readily be carried out manually using standard epifluorescence microscopy. However, there are a number of applications where computer-supported image analysis is helpful or mandatory. These include (but are not limited to) detection of weak signals; image processing, such as optical filtering, image registration, or pseudocolouring; (*Plate 2f*) measurements, such as contour length or intersignal distances; automated analysis; 3D (confocal) microscopy. Laser scanners coupled to a photomultiplier tube or sensitive cameras can be used to capture the data. Camera sensitivity ranges from 10^0–10^{-4} lux for intensified cameras to 10^1–10^{-7} lux for cooled CCD (charge-coupled device) cameras. By comparison, ASA 10 000 film is sensitive over the approximate range 10^1-10^{-4}. Because the CCD camera is efficient and sensitive over a wide range of wavelengths, it has become the most popular instrument for 2D FISH analyses. The hardware cost of such a system will largely be driven by two factors: the grade of CCD array chosen, and the selection of intensified vs. cooled and intensified CCD. CCDs are graded according to the number of defects present, although the number of defects allowed per 'grade' varies with the particular manufacturer. Grade 0 is basically defect-free. This is the most expensive chip, and is probably not required for most gene mapping work, nor for most clinical diagnostic applications. Grade 1 arrays are usually suitable, and in fact many defects can be compensated for in the software. The choice of real-time vs. cooled CCD will depend upon the application. Intensified CCDs achieve video rates, and real-time analysis is more interactive and faster than analysis with the cooled CCD camera. It also lends itself more readily to automated analysis. However, the cooled CCD camera is more sensitive, and thus better suited to extremely low light applications. Confocal (3D) microscopy involves collecting a series of optical sections through the specimen. Computer software is used to reassemble the images, and to 'deconvolve' the images mathematically to remove out-of-focus fluorescence. Laser scanning confocal microscopy removes the majority of out-of-focus fluorescence prior to data collection. Confocal microscopy can be used for 2D FISH analyses, but is really only required for FISH studies of nuclear topography. Commercial hardware and software packages are available to support both 2D and 3D FISH analysis. Furthermore, a number of FISH mapping software programs have been cited in the literature, and are available from the investigator or his/her institution.

References

1. Gall, J. G. and Pardue, M. L. (1969). *Proc. Natl Acad. Sci. USA*, **63**, 378.
2. Harper, M. E. and Saunders, G. F. (1981). *Chromosoma*, **83**, 431.
3. Langer, P. R., Waldrop, A. A., and Ward, D. C. (1981). *Proc. Natl Acad. Sci. USA*, **78**, 6633.
4. Brigati, D. J., Myerson, D., Leary, J. J., *et al.* (1983). *Virology*, **126**, 32.
5. Lo, Y.-M. D., Mehal, W. Z., and Fleming, K. A. (1988). *Nucl. Acids Res.*, **16**, 8719.
6. Weier, H.-U. G., Segraves, R., Pinkel, D., and Gray, J. W. (1990). *Histochem. Cytochem.*, **38**, 421.
7. Kessler, C., Holtke, H. J., Seibl, R., Burg, J., and Muhlegger, K. (1990). *Chem. Hoppe-Seyler*, **371**, 917.
8. Lichter, P., Tang, C. C., Call, K., *et al.* (1990b). *Science*, **247**, 64.
9. Lichter, P., Cremer, T., Tang, C.-J. C., Watkins, P. C., Manuelidis, L., and Ward, D. C. (1988a). *Proc. Natl Acad. Sci. USA*, **85**, 9664.
10. Trask, B. J. (1991). *TIG*, **7**, 149.
11. McNeil, J. A., Johnson, C. V., Carter, K. C., Singer, R. H., and Lawrence, J. B. (1991). *GATA*, **8**, 41.

12. Manuelidis, L. (1985). *Focus*, **7**, 4.
13. Landegent, J. E., Jansen, in de Wal, N., Dirks, R. W., Baas, F., and van der Ploeg, M. (1987). *Hum. Genet.*, **77**, 366.
14. Lichter, P., Cremer, T., Borden, J., Manuelidis, L., and Ward, D. C. (1988b). *Hum. Genet.*, **80**, 224.
15. Pinkel, D., Landegent, J., Collins, C., Fuscoe, J., Segraves, R., Lucas, J., and Gray, J. (1988). *Proc. Natl Acad. Sci. USA*, **85**, 9138.
16. Britten, R. J. and Kohne, D. E. (1968). *Science*, **161**, 529.
17. Cremer, C., Lichter, P., Borden, J., Ward, D. C., and Maneulidis, L. (1988). *Hum. Genet.*, **80**, 235.
18. Jauch, A., Daumer, C., Lichter, P., Murken, J., Schoeder-Kirth, T., and Cremer, T. (1990). *Hum. Genet.*, **85**, 145.
19. Willard, H. F. and Waye, J. S. (1987). *Trends Genet.*, **3**, 192.
20. Lichter, P., Jauch, A., Cremer, T., and Ward, D. C. (1990). In *Molecular genetics of chromosome 21 and Down syndrome* (ed. D. Patterson).
21. Lichter, P., Boyle, A. L., Cremer, T., and Ward, D. C. (1991). *GATA*, **8**, 24.
22. Klinger, K. W., Dackowski, W., Leverone, B., Locke, P., Nass, S. G., Lerner, T., Landes, G., and Shook, D. (1990). *Am. Soc. Hum. Genet.*, **47**, A224.
23. Klinger, K. W., Landes, G., Shook, D., Harvey, R., Lopez, L., Locke, P., Lerner, T., Osathanondh, R., Leverone, B., Houseal, T., Pavelka, K., and Dackowski, W. (1992). *Am. J. Hum. Genet.*, **51**, 55.
24. Fuscoe, J. C., Collins, C. C., Pinkel, D., and Gray, J. W. (1989). *Genomics*, **5**, 100–9.
25. Van Dilla, M. A., Deaven, L. L., Albright, K. L., *et al.* (1986). *Biotechnology*, **4**, 537.
26. Lichter, P., Ledbetter, S., Ledbetter, D., and Ward, D. (1990). *Proc. Natl Acad. Sci. USA*, **87**, 634.
27. Bohlander, S., Espinosa, R., LeBeau, M., Rowley, J., and Diaz, M. (1992). *Genomics*, **13**, 1322.
28. Telenius, H., Pelmear, A. H., Tunnacliff, A., Carter, N. P., Behnel, A., Ferguson-Smith, M. A., Nordenslejold, M., Pfragner, R., and Ponder, B. A. J. (1992). *Genes Chrom. Cancer*, **4**, 257.
29. Bauman, J., Pinkel, D., van der Ploeg, M., and Trask, B. (1989). In *Flow cytogenetics* (ed. J. Gray), pp. 276–301. Academic Press, New York.
30. Waye, J. and Willard, H. (1987). *Nucl. Acids Res.*, **15**, 7549.
31. Fowler, J., Burgoyne, L., Baker, E., Ringenbergs, M., and Callen, D. (1989). *Chromosoma*, **98**, 266.
32. Hutchison, N. J., Langer-Safer, P. R., Ward, D. C, and Hamkalo, B. A. (1982). *J. Cell. Biol.*, **95**, 609.
33. Manuelidis, L., Langer-Safer, P.R., and Ward, D. C. (1982). *J. Cell. Biol.*, **95**, 619.
34. Weier, H.-U. G., Lucas, J. N., Poggensee, M., *et al.* (1991). *Chromosoma*, **100**, 371.
35. Kallioniemi, A., Kallioniemi, O.-P., Sudar, D., Rutovitz, D., Gray, J. W., Waldman, F., and Pinkel, D. (1992). *Science*, **258**, 818.
36. Reid, T., Landes, G., Dackowski, W., Klinger, K., and Ward, D. C. (1992). *Hum. Mol. Genet.*, **1**, 307.
37. Reid, T., Baldini, A., Rand, T. C., and Ward, D. C. (1992). *Proc. Natl Acad. Sci. USA*, **89**, 1388.

38. Dauwerse, J. G., Wiegant, J., Raap, A. K., Breuning, M. H., and van Ommen, G. J. B. (1992). *Hum. Mol. Genet.*, **1**, 593–8.
39. Trask, B. J., Massa, H., Kenwrick, S., and Gitschier, J. (1991). *Am. J. Hum. Genet.*, **48**, 1–15.
40. Brandriff, B., Gordon, L., and Trask, B. (1991). *Genomics*, **10**, 75.
41. Brandriff, B. F., Gordon, L. A., Segraves, R., and Pinkel, D. (1991). *Chromosoma*, **100**, 262.
42. Heng, H. Q., Squire, J., and Tsui, L.-C. (1991). *Am. J. Hum. Genet.*, **49** (Suppl), 368.
43. Wiegant, J., Kalle, W., Mullenders, L., Brooks, S., Hoovers, J. M. N., Dauwerse, J. G., van Omman, G. J. B., and Raap, A. K. (1992). *Hum. Mol. Genet.*, **1**, 587.
44. Houseal, T. W., Dackowski, W. R., Landes, G. M., Mattilla, A., and Klinger, K. W. (1992). *Am. J. Hum. Genet.*, **51**, A27.
45. Wray, W. and Stubblefield, E. (1970). *Exp. Cell. Res.*, **59**, 469.
46. Paulson, J. R. and Laemmli, U.K. (1977). *Cell*, **12**, 817.

3

Non-isotopic detection of DNA in tissues

J. J. O'LEARY, G. BROWNE, M. S. BASHIR, R. J. LANDERS,
M. CROWLEY, I. HEALY, F. A. LEWIS, and C. T. DOYLE

1. Introduction

Over the last three or four decades, new techniques employing immunological and molecular biological assays have had a major impact in histopathology. The field of molecular biology has expanded dramatically with the development of recombinant DNA techniques which make sensitive detection of specific DNA and RNA sequences by molecular hybridization now possible.

All nucleic acid hybridization techniques rely on the annealing of complementary sequences of nucleic acid. Non-isotopic *in situ* hybridization (NISH) differs from other hybridization methods (for example Southern and Northern blotting) in that nucleic acids which are to be identified remain in their cellular environment. Therefore, when one uses a complementary labelled nucleic acid sequence and hybridizes this to a tissue section/smear, the label can subsequently be demonstrated allowing precise cytological localization of the target sequence. NISH has been employed for the identification of genomic, viral, and mRNA sequences in cryostat, paraffin wax, electron microscopic, and chromosomal preparations.

2. Identification of human and viral genes by *in situ* hybridization in tissues

The two types of *in situ* hybridization which are in general use, hybridization to nuclear DNA and hybridization to cellular RNA, conceptually are quite similar, but differ significantly in technical detail. In this chapter, DNA hybridization only will be covered, and only the use of non-isotopic labels will be discussed here. Most *in situ* hybridizations are carried out using preparations that are analysed with the light microscope. However, a technique for hybridization to chromosomes prepared for analysis by electron microscopy has been developed. This technique will not be discussed here and readers should refer to references 1 and 2.

Any human gene sequence may be probed using the NISH technique. The predominant use of NISH in histopathology, however, has been for the identification of viruses in tissue sections. Clinical features may sometimes indicate a diagnosis of viral infection, but, in the past, definitive diagnosis has relied on the demonstration of virus in culture or detection in tissue sections by immunocytochemistry. Even in those laboratories where viral culture is available, it is extremely time consuming and costly, and some viruses cannot be cultured easily, for example human papillomavirus. Immunological detection of virus has proven to be unreliable in many cases, due to the complexities of viral infection. However, by utilizing radiolabelled or hapten-labelled probes of DNA viral sequences, we are now able to identify specific viral infections and to localize the virus to specific cells within infected tissues by *in situ* hybridization. This technique has been used to identify many viral infections and these applications have been reported previously. The most commonly studied is human papillomavirus (HPV), which has now been identified in anal, cervical, and laryngeal lesions (3–5). Non-homology between HPV types (of which there are now over 60) exists. This enables specific virus typing by *in situ* hybridization to be carried out, provided that adequate stringency is applied when using the method. Herpes viruses have also been studied using *in situ* hybridization, and these include cytomegalovirus infection (6), herpes simplex virus type I in brain tissue (7), and Epstein–Barr virus in Hodgkin's disease (8), in primary brain lymphoma (9), and cervical carcinoma (10). With the sophistication of NISH techniques, even low copy viral numbers in infected cells can now be detected. Episomal viral infections are easily detected using most NISH protocols.

2.1 Selection of probes for *in situ* hybridization

Recombinant DNA technology now provides the opportunity to use DNA or RNA probes to study any desired sequence. Furthermore, one can choose between single-stranded and double-stranded probes. Which type of probe should be used in *in situ* hybridization protocols depends on the particular use of the probe. In terms of viral identification, viruses are classified and characterized as being DNA or RNA types. DNA viruses may be double- or single-stranded, whilst RNA viruses contain sense or antisense forms. In addition, some RNA viruses are known to replicate via an intermediate double-stranded cDNA step. Probes to DNA viruses are constructed as recombinant plasmids or cosmids, containing whole viral DNA or restriction enzyme fragments of specific viral sequences. Many viral and human probes are now commercially available, and may be obtained from the originating laboratory or from the American Type Culture Collection (ATCC). Double-stranded probes, if randomly sheared, form networks on the cytological hybrid and so increase the hybridization signal. However, double-stranded probes can also anneal in solution and thus reduce the concentration of probe available for reaction with the cytological preparation. Recently, human and viral probes have been constructed synthetically using oligonucleotide synthesizers. These oligonucleotides sequences are complementary to DNA or RNA viruses and can be easily labelled and used as probes, but their use in conjunction with non-isotopic detection systems has been limited. This is due to a lack of sensitivity which results from a lower amount of label available on these short sequences. To use a synthetic oligonucleotide successfully in non-isotopic methods it has been found that probe cocktails containing up to 16 different sequences are necessary.

2.2 Probe label selection

The general use of ISH was initially hindered by problems associated with probe preparation. Initially probes were prepared using crude isolated nucleic acids. In addition, only cumbersome, expensive, and dangerous radioactive labels could be used. These difficulties were largely overcome by recombinant DNA technology. The development of a biotinylated nucleotide analogue, biotin-11-dUTP (Enzo Diagnostics Inc.), and the development of the DNA labelling method of nick-translation, have greatly facilitated the use of NISH. *In situ* hybridization using biotinylated viral probes was described using immunocytochemical detection, but the development of a technique using a biotin/streptavidin/alkaline phosphatase sandwich detection system proved to be more sensitive (11). Subsequently, multiple sandwich immunohistochemical techniques have become available (12–15). However, problems exist when using biotin, as many tissues, for example liver, small bowel, and endometrium, have high levels of endogenous biotin. This obviously interferes with the detection of biotinylated hybridization products. Therefore, alternative label molecules have been used for the detection of DNA sequences. Aminoacetylfluorene- and mercury-labelled probes can detect nucleic acids in fresh isolates or cultured cells by fluorescent methods. In general, fluorescence detection systems are not readily applicable to archival paraffin wax embedded sections because of tissue autofluorescence. Aminoacetylfluorene is also carcinogenic and mercury is toxic. Among the alternative labels to biotin, digoxigenin (Boehringer–Mannheim) — a derivative of the cardiac glycoside digoxin — conjugated to deoxyuridine triphosphate, is the most useful. Many immunohistochemical techniques for the detection of digoxigenin are now available and will be discussed later. Most DNA probes (especially viral) contain long sequences of nucleic acid, and consequently we find that nick-translation in the presence of a biotinylated/digoxigenin deoxynucleotide is the most effective method for probe labelling with these labels. Nick-translation has the one advantage of producing large quantities of labelled probe, which can readily be stored at −20 °C for up to two years without any significant deterioration. In addition, the nicking procedure of the DNA probe during the labelling reaction results in a labelled probe with sequence lengths that are most appropriate for *in situ* hybridization (300–800 bp). Also, it appears useful to label the entire recombinant plasmid so that the probe can be used with 'built-in carrier DNA', that enhances sequence strength by the formation of networks with specific sequences. Random primer labelling methods require pre-digestion of a recombinant plasmid to a linear form or excision and purification of the viral sequence. Among the biotin deoxynucleotides available, we have identified biotin-11-dUTP as a nucleotide that produces biotin probes at optimum detection sensitivities. The labelling methods for DNA probes are presented in Chapter 1. Purification of labelled probes can be achieved using the spun-column purification technique which removes unincorporated nucleotides from labelled probes and thereby avoids high levels of background staining.

3. Preparation of sections for *in situ* analysis

3.1 Slide preparation

Human DNA sequences and DNA viral sequences can be detected by *in situ* hybridization in cell smears, cryostat sections, fresh frozen tissue, and fixed paraffin wax embedded tissue. In all *in situ* hybridization applications we recommend the use of either single rectangle well slides (type PH 106 from C. A. Hendley) or four spot multi-well slides, diameter 12 mm (type PH 005, C. A. Hendley). The wells formed by the thin Teflon coating on these slides localize reagents so that a minimum quantity can be employed. In addition, the Teflon border allows for clearance between the section and the coverslip during hybridization procedures and minimizes section loss. To ensure maximum section adhesion and to minimize section loss during the harsh pre-treatments required, slides should be pre-coated with 3-amino-propyl-triethoxysilane (see *Protocol 1*). Gelatin and poly-L-lysine are also routinely employed as slide-coating agents. These reagents, however, tend to form smears which can bind many of the reagents used in hybridization procedures and many routine histological stains. In addition, bacterial growth on slides occurs with these coatings and this may lead to spurious high background staining.

Protocol 1. Coating slides with aminoalkylsilane

Reagents

• 2% (v/v) Decon 90

• 2% (v/v) 3-amino-propyl-triethoxysilane in acetone

Method

1. Immerse the slides in 2% (v/v) Decon 90, in warm water for 30 min.
2. Wash thoroughly with water to remove the detergent.

---

3. Drain off excess water and immerse in acetone for 1–2 min.[a]

4. Drain off acetone and immerse in 2% amino-propyl-triethoxysilane in acetone for 5 min.

5. Drain off excess solution and wash in running water for 1–2 min.

6. Drain off excess water and allow the slides to dry overnight at room temperature. The slides can be stored at room temperature until needed.

[a] Perform steps 3 and 4 in a fume hood.

3.2 Preparation of cell smears

3.2.1 Cultured cells

(a) Wash the cultured cells at least three times with phosphate buffered saline (PBS), pH 7.4, at 4°C to remove all traces of culture medium.

(b) Resuspend the cells at a concentration of 2×10^6 cells per ml of PBS and pipette 50 μl into the centre of a pre-coated well slide. Allow the cells to settle for 10 min, shake off excess suspension, and fix immediately as described in *Protocol 3*.

3.2.2 Cervical smears

Incubate routine cervical smear slides in methanol/acetic acid (3:1, v/v) for 10 min at room temperature and fix immediately as in *Protocol 4*.

3.3 Preparation of cryostat sections of fresh frozen tissue

(a) Cut 5 μm sections.

(b) Place the sections on pre-coated well slides and wash the sections twice in PBS, pH 7.4, at 4°C to remove all traces of freezing medium (OCT compound). This material binds many of the detection reagents which results in very high background.

(c) Fix sections immediately as described in *Protocol 2*.

3.4 Preparation of fixed paraffin-embedded tissues

The permeability of fixed tissue to DNA probes is dependent on the particular type of fixative used. Precipitating fixatives, such as acetone or ethanol/ acetic acid mixtures, allow probe penetration with limited pre-treatment of the tissue sections. However, such fixatives permit loss of small target nucleic acids and usually yield poor morphology after hybridization. The 'cross-linking' fixatives, such as formalin or paraformaldehyde, allow probe penetration only after considerable pre-treatment of tissue sections. The use of these fixatives, however, favours complete preservation of morphology and nucleic acids, despite the harshness of the pre-treatment protocols required in the hybridization method. The use of highly cross-linking fixatives, such as glutaraldehyde or mercuric chloride modified formalin, should be avoided, as they render tissue virtually impermeable to DNA probes, which results in poor sensitivity of detection and poor morphology due to the extensive pre-treatment protocols that are required. Similarly, Bouin's fixed tissue is unsuitable for *in situ* hybridization due to the extensive denaturing of the nucleic acids by the picric acid in the fixative. Routinely, therefore, tissue should be fixed in 10% (v/v) aqueous formalin or 4% (v/v) formaldehyde solution prior to paraffin wax embedding. *Protocol 2* describes the preparation of paraffin-embedded tissues.

Protocol 2. Preparation of fixed paraffin embedded tissues

Reagents

- 10% (v/v) aqueous formalin *or*
- 4% (v/v) formaldehyde solution
- xylene
- absolute alcohol
- graded alcohol series to water

Method

1. Fix tissue in 10% (v/v) aqueous formalin *or* 4% (v/v) formaldehyde solution prior to paraffin wax embedding.
2. Cut 5 μm sections, place on well slides, and incubate on a hot plate for 24–48 h in order to achieve maximum section adhesion on the slides. This step prevents loss of sections during the pre-treatment steps.
3. Section de-waxing should then be carried out by entirely immersing the slides in the following solutions
   - xylene at 37°C for 30 min;
   - xylene at room temperature for 10 min;
   - absolute alcohol at room temperature for a further 10 min;
   - fresh absolute alcohol.
4. Rehydrate through a graded alcohol series to water over a 10 min period, prior to performing the pre-treatment steps as described in *Protocol 5*.

---

4. Pre-treatments required before hybridization

Because of the limited fixation employed for cell smears and cryostat sections of frozen tissue, pre-treatment required for these sections is different to that required for sections cut from paraffin wax embedded sections.

4.1 Pre-treatment of fresh culture cells and cryostat sections

The aim of any pre-treatment protocol is to permeabilize the tissue to allow the target nucleic acid sequence to be accessed by the DNA probe used. Incubation in detergent solutions produces sufficient permeabilization of fresh culture cells and cell smears. A brief incubation in acetic acid is used to destroy endogenous alkaline phosphatase activity. This step is effective for the removal of liver type alkaline phosphatase, but is usually less effective in destroying intestinal and placental types of the enzyme. Incubation of sections in glycerol solution considerably improves the hybridization efficiency and appears to obviate the need for a pre-hybridization step. Glycerol may act by excluding water within cells or may enable the rapid access of probe thereby increasing the initial rate of hybridization. These techniques are described in *Protocol 3*.

Protocol 3. Pre-treatment of fresh cultured cells, cell smears, and cryostat sections

Reagents

- phosphate buffered saline (PBS), pH 7.4
- 4% paraformaldehyde in PBS. Just before use dissolve 12 g of paraformaldehyde in 300 ml PBS by heating to near boiling point. Cool rapidly on ice to room temperature
- 0.1 M Tris–HCl, pH 7.2

- 0.1 M Tris–HCl, pH 7.2, 0.25% Triton-X-100, 0.25% Nonidet P-40
- 2 × SSC, pH 7.4: 0.3 M sodium chloride, 30 mM sodium citrate
- aqueous 20% glycerol
- acetic acid

Method

1. Fix the cells in 4% paraformaldehyde in PBS for 30 min.

2. Rinse the slides in PBS for 5 min.

3. Immediately extract in 0.1 M Tris–HCl, pH 7.2 containing 0.25% Triton X-100 and 0.25% Nonidet P-40, twice for 5 min each time.

4. Wash the slides in 0.1 M Tris–HCl, pH 7.2, twice for 5 min each time.

5. Immerse the slides in 20% (v/v) aqueous acetic acid at 4°C for 15 sec.

6. Wash the slides in 0.1 M Tris–HCl, pH 7.2, three times for 5 min each time.

7. Incubate the slides in aqueous 20% glycerol for 30 min at room temperature.

8. Rinse with 0.1 M Tris–HCl, pH 7.2

9. Immerse the slides in 2 × SSC for 10 min.

4.2 Pre-treatment of cervical smears

A specialized method is presented in *Protocol 4* for the pre-treatment of cervical smears, for use with a peroxidase detection system.

Protocol 4. Pre-treatment of cervical smears (15)

Reagents

- 4% paraformaldehyde in PBS
- 0.2% (w/v) glycine

- 0.1% (w/v) sodium azide/0.3% hydrogen peroxide (v/v)
- proteinase K solution (1 μg/ml) in PBS

Method

1. Fix in 4% paraformaldehyde in PBS for 15 min at room temperature.
2. Wash in PBS containing 0.2% (w/v) glycine for 5 min.
3. Rinse the slides in PBS.
4. Immerse the slides in 0.1% (w/v) sodium azide containing 0.3% hydrogen peroxide (v/v) for 10 min. This is to abolish endogenous peroxidase activity.
5. Wash slides in PBS for 5 min.
6. Incubate the slides in proteinase K solution for 15 min at 37°C.
7. Wash in PBS for 5 min.
8. Immerse the slides in 4% paraformaldehyde for 5 min.
9. Wash in 0.2% PBS/glycine for 5 min.
10. Wash in PBS for 5 min.
11. Air dry at 37°C.

4.3 Pre-treatment of formalin-fixed paraffin-embedded tissues

The pre-treatment of formalin-fixed paraffin sections is described in *Protocol 5*. In this protocol we use a dilute hydrochloric acid hydrolysis step, the significance of which is not clear. However, omission of this step significantly reduces the hybridization signal. One possible explanation for this is that under conditions of mild acid hydrolysis, some limited depurination of nucleic acids occur. The hydrochloric acid may also partially solubilize the highly cross-linked acidic nuclear proteins (which occur in fixed tissues) and thus enable easier access of the probe. An incubation with detergent solution alone is insufficient for effective permeabilization of the cells, but this step should be included to partially permeabilize the cell membranes. For complete permeabilization, however, incubation with a protease enzyme is essential. Protease enzymes are effective in degrading many of the cross-linked proteins that are formed as a result of fixation and, therefore, allow easier access of probe to the DNA. We find that proteinase K from Life Technologies Ltd is the most effective proteinase for performing this task, without destroying overall cellular architecture. The concentration of proteinase K used is critical if morphology is to be preserved, and it is clearly tissue dependent. The suitable concentration for the particular tissue under investigation must be determined empirically by titration using a positive control for hybridization. Following protease treatment, sections are incubated in acetic acid to reduce the level of endogenous alkaline phosphatase in the tissues. The harsher pre-treatments of the protocol described partially reverse the effects of the initial fixation, in order to avoid loss of nucleic acid sequences from the tissue during the remaining steps of the hybridization protocol. Sections are fixed briefly in paraformaldehyde before dehydration with ethanol, so as to ensure maximum preservation of tissue morphology.

Protocol 5. Pre-treatment of paraffin embedded tissue

Reagents

- PBS
- 20 mM HCl
- 0.01% Triton-X-100 in PBS
- proteinase K buffer: 50 mM Tris–HCl, pH 7.4, 5 mM EDTA

- proteinase K
- 2 mg/ml glycine in PBS
- 20% acetic acid
- 4% paraformaldehyde in PBS (*Protocol 2*)
- alcohol series: 50%, 70%, 95%, 100%

Method 1

1. Immerse the prepared slides (see Section 3.4) in 20 mM HCl for 10 min.
2. Wash twice in PBS for 5 min each.
3. Extract with 0.01% Triton X-100 in PBS for 3 min.
4. Wash twice in PBS for 5 min each.
5. Incubate the slides in pre-warmed proteinase K buffer at 37°C for 10 min.
6. Incubate the slides with proteinase K at 37°C for 10–20 min. The concentration of proteinase K (0.1–5 mg/ml) and the time of digestion will vary with the tissue under investigation and can be calculated from trial experiments.
7. Wash the slides in two changes of PBS containing 2 mg/ml of glycine for 5 min each. This stops the action of proteinase K.
8. Immerse the slides in aqueous 20% acetic acid at 4°C for 15 sec.
9. Wash in two changes of PBS for 10 min each.
10. Post-fix in 4% paraformaldehyde in PBS for 5 min.

Protocol 5. *Continued*

11. Wash in PBS for 5 min.
12. Dehydrate through graded alcohol series.

Method 2

(This is a quicker protocol than Method 1, but cellular architecture preservation is not as good.)

1. Spot proteinase K in PBS (0.1–5 mg/ml) on to the prepared tissue section (see Section 3.4) and place in either a pre-heated oven at 37°C for 10 min *or* in a Terasaki plate floating in a waterbath at 37°C for 15 min.
2. Wash in distilled water and air-dry at 75°C.

5. Hybridization of probes to pre-treated tissue sections

Following pre-treatment, the methods for hybridization and detection of the hybridization signal are essentially the same for all types of tissue section.

5.1 Hybridization method

Biotinylated and digoxigenin labelled DNA probes (Chapter 1) are usually prepared at a concentration of 200 ng/ml in one of the hybridization buffers described in *Table 1*. It is usually preferable to use as simple a hybridization buffer as possible, and this is usually adequate for producing successful hybridization at temperatures between 37 and 42°C with overnight incubation. Hybridization buffer 2 (*Table 1*) is an example of this. The milk powder present in this buffer successfully blocks non-specific probe binding sites, which obviates the need for complex mixtures containing phenol, bovine serum albumin (BSA), and polyvinylpyrrolidone (PVP). In addition, there appears to be no significant advantage in incorporating a carrier DNA, such as single-stranded DNA (e.g. salmon sperm), into hybridization mixtures. Hybridization buffers that contain dextran sulfate have a high viscosity. This usually produces a high surface tension on contact with a glass surface. Therefore, if glass coverslips are used to cover tissue sections during hybridization reactions, they are subsequently difficult to remove as the surface tension which is induced between the glass and the buffer produces enough suction to lift the section from the slide. For this reason, we recommend the use of gel bond film (FMC Corp.), cut to coverslip size to cover the sections. This material is a pliable plastic and is, therefore, easily removed following hybridization. It also has a hydrophilic and hydrophobic surface. The hydrophobic surface is usually placed face downwards towards the tissue section

---

Table 1. Hybridization solutions for DNA probes

(a) *Hybridization buffer 1*
- 50 mM Tris–HCl pH 7.2, 10% dextran sulfate, 2 × SSC, 2 × Denhardts, 200 μg/ml single-stranded salmon sperm sheared to 200 bp, 50% formamide.
  Make up a stock solution of 2 × hybridization buffer without formamide. Prior to use add an equal volume of formamide.
  100 × Denhardts is 2% BSA, 2% Ficoll, 2% polyvinylpyrrolidine

(b) *Hybridization buffer 2*
- 2 × SSC, 5% dextran sulfate, 0.2% (w/v) dried milk powder (pure, containing no vegetable extracts), 50% formamide.
This simplified buffer, also known as 'blotto buffer', is easy to make and is an alternative to buffer 1 described above giving results comparable to or better than buffer 1.

(c) *Hybridization buffer 3*
- 50% formamide, 5% dextran sulfate, 2 × SSC, 50 mM Tris–HCl, pH 7.4, 0.1% (w/v) sodium pyrophosphate, 0.2% (w/v) polyvinyl pyrrolidone (mol. wt 40 000) 0.2% (w/v) Ficoll (mol. wt 400 000), 5 mM EDTA, 200 ng/ml sheared human DNA

  This buffer is particularly useful when cocktails of genomic probes are used for the analysis of routine cervical smears for the detection of HPV 6, 11, 16, 18, 31, 33, etc.

(d) *Hybridization buffers for oligo probes*
- 2 × SSC, 5 × dextran sulfate, 0.2% dried milk powder

---

which results in no surface tension with the buffer. Gel bond is sealed in place with nail varnish to prevent leakage out of the probe and leakage in of moisture during hybridization. Alternatively, rubber cement *or* agarose can be used. Before hybridization (either overnight or for 4 h) at the desired temperature, both cellular DNA and probe DNA (if double-stranded) must be simultaneously denatured by heating the slides to 90–95 °C for ten minutes. This is accomplished by placing the slides on a pre-heated baking tray in an oven, set at the desired temperature. Both the temperature and timing of this step are critical if tissue morphology is to be preserved. Incubation of the slides at 37–42 °C overnight results in high hybridization efficiency, particularly with DNA virus probes, and is stringent enough to produce specific hybridization with little or no cross-hybridization between viral types.

Protocol 6. Hybridization of DNA probes to pre-treated tissue sections

Equipment and reagents

- hybridization buffer (see *Table 1*)
- probe
- single well slides (C. A. Hendley PH 106)

- *or* double well slides (C. A. Hendley PH 005)
  —see Section 3.1

Protocol 6. *Continued*

Method

1. Apply 75 µl of the appropriate probe in the selected hybridization buffer (1, 2, 3, or 4) to the centre of the well of the glass slide. For multiwell slides, apply 8–10 µl of probe to the centre of the well.[a]

2. Cut gel bond films to coverslip size and place hydrophobic side down, over each section. Seal the gel bond in place with nail varnish.[b]

3. Place the slide on to a pre-heated baking tray and incubate at 90–95°C for 10 min.

4. Transfer the slides to a humidified box and incubate slides at 37–42°C overnight or at 42°C for 2 h.

---

[a] For cervical smears, use hybridization buffer 3 (*Table 1*) and a cocktail of probes (HPV 6, 11, 16, 18, 31, 33) at a concentration of 2 ng/µl of each probe.
[b] For multiwell slides, a plastic coverslip may be used but it is advisable not to seal with nail varnish.

5.1.1 Hybridization controls

It is important to include hybridization controls that: (a) indicate that hybridization has occurred, (b) show no cross-hybridization with vector sequences used, and (c) indicate the level of background development due to non-specific effects. Hybridization controls must be included for each specimen under investigation. Include biotinylated/digoxigenin labelled genomic DNA extracted from, for example, placenta, for use as a total DNA probe; i.e. a positive control that should hybridize to every nucleus in the tissue section. If no result is obtained with this probe then hybridization has failed. A negative control probe can be biotinylated/digoxigenin labelled plasmid such as pBR322. If a positive signal is obtained with this probe, then cross-hybridization with vector sequences has occurred. Hybridization buffer alone should provide a negative control that indicates the level of non-specific signal after development. It is also useful to include a non-related probe, i.e. to an unrelated genetic sequence, as an internal control for the reaction. Positive tissue controls include cell lines and positive material containing the target sequence. Negative tissue controls are cell lines that do not contain the target sequence or tissues negative for the particular target.

5.2 Post-hybridization washes

Post-hybridization washes (*Protocol 7*) are more efficiently performed with agitation of the washing buffer. This is accomplished simply by inverting a Coplin jar lid into the bottom of a glass staining dish and placing a magnetic follower in the upturned lid. Fill the dish with wash buffer and immerse the staining rack into the buffer so that it rests on the edge of the Coplin jar lid.

The buffer can now be stirred vigorously on a magnetic stirrer without impairment of the magnet's motion by the staining rack or slides.

5.2.1 Washing with increasing stringency

The object of the washing protocol is to remove excess and non-specifically bound probe and to melt away any mismatched hybrids that may have formed during hybridization. This is achieved by increasing the stringency of washing conditions with decreasing salt concentration and increasing temperature. The post-hybridization washes can be adapted to the required stringency by varying the SSC concentrations and the temperature of washing. Two sequential washing procedures are described in *Protocol 7*, which produce medium to high stringency, sufficient for the detection of specific hybridizations using DNA probes.

Protocol 7. Post-hybridization washes

Reagents

- 4 × SSC: 0.6 M sodium chloride, 60 mM sodium citrate
- 2 × SSC
- 0.2 × SSC

Method 1

1. Remove the gel bond coverslips with a scalpel blade.
2. Wash the slides in SSC with agitation according to the following protocol:
   - 2 × SSC at room temperature for 10 min.
   - 2 × SSC at 60 °C for 20 min.
   - 0.2 × SSC at room temperature for 10 min.
   - 0.2 × SSC at 42 °C for 20 min.
   - 0.1 × SSC at room temperature for 10 min.
   - 2 × SSC at room temperature for 1–2 min.

or

Method 2

1. Wash slides in 4 × SSC at room temperature three times for 5 min each.

---

6. Detection of hybridization signal

Many detection techniques are now available for the detection of biotinylated and digoxigenin labelled human and viral DNA probes. For convenience, these will be described as one-step, two-step, three-step and five-step pro-

---

Table 2. Solutions required for non-isotopic detection of hybridization signal

(a) *Buffer 1*: 0.1 M Tris–HCl, pH 7.5, 0.1 M NaCl, 2 mM $MgCl_2$, 0.05% Triton-X-100

(b) *Buffer 2*: 0.1 M Tris–HCl, pH 9.5, 0.1 M NaCl, 50 mM $MgCl_2$

(c) *Tris buffered saline* (TBS): 50 mM Tris–HCl, 100 mM NaCl, pH 7.2

(d) *Blocking reagent* (TBT): TBS + 3% BSA, 0.5% Triton-X-100

(e) *Alkaline phosphatase system* (NBT/BCIP)
  • nitro-blue tetrazolium (NBT), 75 mg/ml in dimethylsulphoxide (DMSO);
  • bromo-chloro-3-indoyl phosphate (BCIP), 50 mg/ml in DMSO;
  • just prior to use combine 16.5 μl of NBT solution with buffer 2; mix gently by inversion;
  • add 12.5 μl of BCIP and mix gently again, store in the dark.

(f) *Peroxidase system* (AEC)
  • H_2O_2 30% (v/v) (Merck);
  • 3-amino-9-ethyl carbazole (AEC) (Sigma);
  • 20 mmol acetate buffer, pH 5.0–5.2;
  • DMSO;
  • dissolve 2 mg AEC in in 1.2 ml DMSO in a glass tube; add 10 ml of acetate buffer and 0.8 ml of H_2O_2; alternatively, the AEC substrate kit from Zymed (USA) can be used.

---

cedures, depending on the number of reagents used and the steps involved in the detection protocol. In addition, the use of immunogold silver staining (IGSS) for the detection of biotinylated probes will also be described (*Protocol 11*). The solutions in *Table 2* are required for the detection of hybridization signals using the protocols mentioned below.

6.1 Detection of biotinylated probes

Protocols 8, 9, 10 and *11* describe different detection protocols for biotinylated probes. The choice of detection system depends on the level of sensitivity required (see *Figure 1*).

Protocol 8. One-step procedure

Reagents

• TBT (*Table 2*)
• TBS (*Table 2*)
• avidin alkaline phosphatase (DAKO, UK)

• *or* avidin peroxidase (DAKO)
• colour detection reagents (*Table 2*)

Method

1. Immerse the slides from *Protocol 7* (Method 2) in TBT (blocking reagent) at room temperature for 10 min.

2. Transfer the slides to a slide incubation tray. Transfer small numbers of slides at a time to prevent sections drying out.

3. Incubate slides in avidin alkaline phosphatase *or* avidin peroxidase, diluted 1/100 in TBT.

4. Remove unbound conjugate by washing for 5 min twice in TBS.

5. Incubate slides in NBT/BCIP *or* AEC development reagent for 10–30 min and monitor colour development.

6. Terminate the colour development reaction by washing in distilled water for 5 min.

Protocol 9. Two step procedure

Reagents

- buffer 1 + 5% BSA (*Table 2*)
- buffer 2 (*Table 2*)
- avidin DN (Vector)
- biotinylated alkaline phosphatase (Vector)
- NBT/BCIP reagents (*Table 2*)
- PBS

Method

1. Transfer the slides from *Protocol 7* (Method 1) into buffer 1 containing 5% (w/v) BSA and incubate at room temperature for a minimum of 30 min.

2. Wipe excess buffer from the slides and transfer them to a slide incubation tray. Transfer small numbers of slides at a time to try and prevent sections from drying out.

3. Add a few drops of buffer 1 containing avidin DN at a concentration of 10 μl/ml and incubate at room temperature for 10 min.

4. Wash the slides with agitation with two changes of buffer 1 for 10 min each.

5. Return the slides to the incubation tray and add a few drops of buffer 1 containing biotinylated alkaline phosphatase at a concentration of 10 μl/ml to each section. Incubate at room temperature for 10 min.

6. Wash the slides with agitation twice with buffer 1 for 10 min each.

7. Transfer the slides into buffer 2 and allow to equilibrate for 30 min.

8. Return the slides to the incubation tray and cover the sections with NBT/BCIP development reagent. Monitor the development of the colour after 5 min and then continuously until development looks complete.

9. Terminate the reaction by immersing the slides in PBS or distilled water for 5 min.

J. J. O'Leary et al.

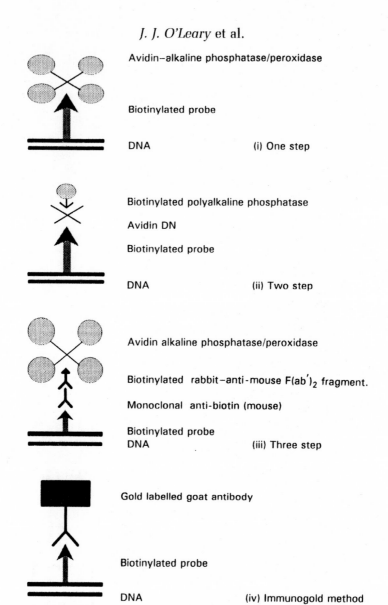

Avidin–alkaline phosphatase/peroxidase

Biotinylated probe

DNA   (i) One step

Biotinylated polyalkaline phosphatase

Avidin DN

Biotinylated probe

DNA   (ii) Two step

Avidin alkaline phosphatase/peroxidase

Biotinylated rabbit–anti-mouse F(ab')$_2$ fragment.

Monoclonal anti-biotin (mouse)

Biotinylated probe
DNA   (iii) Three step

Gold labelled goat antibody

Biotinylated probe

DNA   (iv) Immunogold method

Figure 1. Schematic diagram for detection methods of biotinylated probes.

Protocol 10. Three-step procedure

Reagents

- TBT (*Table 2*)
- TBS (*Table 2*)
- non-fat dried milk

- monoclonal mouse anti-biotin (DAKO) in TBT
- biotinylated rabbit–anti-mouse F(ab')$_2$ fragment (DAKO)

3: Non-isotopic detection of DNA in tissues

- avidin alkaline phosphatase (Vector)
- or streptavidin peroxidase (DAKO)
- colour detection reagents (Table 2)

Method

1. Immerse the slides from *Protocol 7* (Method 2) in TBT (blocking reagent) at room temperature for a minimum of 30 min.

2. Transfer the slides to a slide incubation tray and incubate in monoclonal mouse antibiotin diluted 1/50 in TBT.

3. Wash the slides twice in TBS for 10 min each.

4. Then incubate slides in biotinylated rabbit–anti-mouse F(ab')$_2$ fragment for 30 min.

5. Wash twice in TBS for 5 min.

6. Incubate slides in either avidin alkaline phosphatase diluted 1/50 in TBT, or streptavidin peroxidase diluted 1/100 in TBT containing 5% non-fat milk.

7. Wash in TBS for 5 min.

8. Incubate slides in either NBT/BCIP development reagent or AEC for 10–30 min, as appropriate.

9. Terminate the colour development reaction by washing in distilled water for 5 min.

Protocol 11. Detection of biotin labelled probe using gold-labelled goat anti-biotin

Reagents

- Lugol's iodine
- TBS (Table 2)
- TBT (Table 2)
- TBS 2: 0.8% (w/v) BSA, 0.1% (w/v) gelatin, 5% (v/v) normal swine serum, 2 mM sodium azide
- goat antibody solution 1:10 (Auroprobe) in TBS 2
- silver developing reagents:
  - gum acacia 500 g/l of citrate buffer
  - citrate buffer, pH 3.5:2–3.5 g trisodium citrate · 2H$_2$O, 25.5 g citric acid in 100 ml dH$_2$O
  - hydroquinone 0.85 g/15 ml
  - silver lactate 0.11 g/15 ml
- 2.5% aqueous sodium thiosulfate
- alcohol series

Method

1. Transfer the slides from *Protocol 7* (Method 1), into Lugol's iodine for 2 min. Rinse with TBS, decolourize in 2.5% (w/v) aqueous sodium thiosulfate and wash in TBS twice for 5 min each.

2. Immerse the slides in TBS 2 for 20 min.

3. Remove excess buffer and transfer slides from the incubation tray. Apply 1 ml of gold-labelled goat-antibody solution.

Protocol 11. *Continued*

4. Wash sections in TBS with agitation twice for 5 min each.

5. In a dark room prepare a silver developing solution by adding hydro-quinone and silver lactate to the gum acacia/citrate buffer solution.

6. Immerse the slides in the developing solution in the dark (use an S902 or F904 safelight) until the sections appear optimally developed, when viewed by light microscopy (3–10 min).

7. Wash in running tap water for 5 min.

8. Fix in 2.5% (w/v) aqueous sodium thiosulfate for 3 min.

9. Wash in running tap water for 1 min, counterstain as required, de-hydrate, clear and mount with synthetic resin.

6.2 Detection of digoxigenin labelled probes

Protocols 12, 13, and *14* describe detection techniques for digoxigenin labelled probes. The choice of detection system depends on the level of sensitivity required — the greater the number of steps, the more sensitive the technique (see *Figure 2*).

Protocol 12. One-step detection

Reagents

- TBT (*Table 2*)
- TBS (*Table 2*)

- Alkaline phosphatase conjugated anti-digoxigenin (Boehringer–Mannheim)
- NBT/BCIP detection reagents (*Table 2*)

Method

1. Immerse the slides from *Protocol 7* (Method 2) in TBT (blocking reagent) at room temperature for 10 min.

2. Transfer the slides to a slide incubation tray.

3. Incubate sections in alkaline phosphatase conjugated anti-digoxigenin diluted 1/600 in TBT.

4. Wash in TBS for 5 min, twice.

5. Develop the signal using NBT/BCIP development reagent for 10–30 min and monitor colour development.

6. Terminate the colour development reaction by washing in distilled water for 5 min.

3: Non-isotopic detection of DNA in tissues

Detection of digoxigenin labelled probes

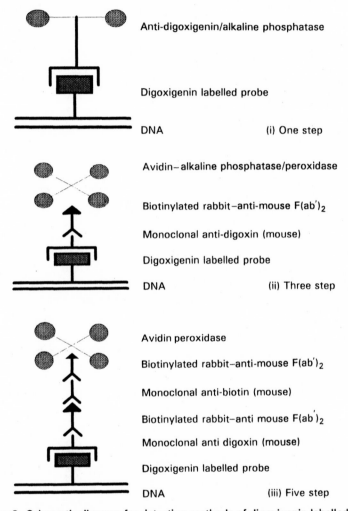

Figure 2. Schematic diagram for detection methods of digoxigenin labelled probes.

Protocol 13. Three-step detection (14)

Reagents

- TBT (*Table 2*)
- TBS (*Table 2*)
- Mouse monoclonal anti-digoxin (Sigma)
- Biotinylated rabbit–anti-mouse F(ab')₂ (DAKO) fragment

- avidin alkaline phosphatase or avidin peroxidase
- colour reagents (*Table 2*)

Protocol 13. *Continued*

Method

1. Immerse the slides from *Protocol 7* (Method 2) in TBT at room temperature for 10 min.

2. Transfer the slides to a slide incubation tray.

3. Incubate slides in monoclonal anti-digoxin diluted in 1/10 000 in TBT.

4. Wash in TBS twice for 5 min each.

5. Incubate in biotinylated rabbit—anti-mouse F(ab')$_2$ fragment diluted 1/200 in TBT.

6. Wash in TBS twice for 5 min each.

7. Incubate the slides in avidin—alkaline phosphatase, diluted 1/50 in TBT or avidin-peroxidase diluted 1/75 in TBT containing 5% (w/v) non-fat milk.

8. Wash in TBS for 5 min.

9. Incubate the slides in NBT/BCIP *or* AEC development reagent as appropriate (*Table 2*).

10. Terminate the colour development reaction by washing in distilled water for 5 min.

Protocol 14. Five-step detection (15)

Reagents

- TBT (*Table 2*)
- TBS (*Table 2*)
- monoclonal anti-biotin (DAKO)

- biotinylated rabbit—anti-mouse F(ab')$_2$ fragment (DAKO)
- avidin peroxidase
- AEC colour reagents (*Table 2*)

Method

1. Follow Steps 1—5 of *Protocol 13* then incubate slides in monoclonal anti-biotin diluted 1/50 in TBT.

2. Wash the slides in TBS for 5 min.

3. Incubate the slides in biotinylated rabbit—anti-mouse F(ab')$_2$ fragment for 10 min.

4. Wash in TBS for 5 min.

5. Incubate the slides in avidin peroxidase diluted 1/75 in TBT containing 5% (w/v) non-fat milk.

6. Incubate the slides in the AEC development reagent for 10–30 min and monitor colour development.

7. Terminate the colour development reaction by washing in distilled water for 5 min.

6.3 Simultaneous detection of biotin and digoxigenin labelled probes (13)

The simultaneous detection of two different nucleic acids can be achieved in the one tissue section using a combined application of biotin and digoxigenin labelled probes to their respective targets in the tissue section (*Protocol 15*).

Protocol 15. Dual probe detection (13)

Reagents

- TBT (*Table 2*)
- TBS (*Table 2*)
- streptavidin/peroxidase conjugate (DAKO)
- alkaline phosphatase conjugated anti-digoxigenin (Boehringer–Mannheim)
- colour detection reagents (*Table 2*)

Method

1. Following post-hybridization washes (*Protocol 7*) incubate the slides at room temperature for 30 min in a mixture of streptavidin/peroxidase conjugate, diluted 1/100 in TBT, and alkaline phosphatase conjugated anti-digoxigenin, diluted 1/600 in TBT.

2. Remove unbound conjugate by washing twice in TBS for 5 min each time.

3. Incubate in AEC development reagent for 30 min at room temperature.

4. Terminate the reaction by thoroughly washing in TBS.

5. Wash in buffer 2 for 10 min.

6. Incubate in NBT/BCIP development reagent for 20–40 min.

7. Terminate the reaction by washing in distilled water for 5 min.

8. Air-dry the slides at 42°C and mount in glycerol jelly.

6.4 Counterstain

Following detection of the hybridization product, the section should be counter-stained with an aqueous stain. In the case of alkaline phosphatase (NBT/BCIP) detection, counterstain with 2% (w/v) methyl green and mount in glycerol jelly. All slides detected using AEC should be counterstained progressively with haematoxylin for approximately 10–15 sec and then mounted in glycerol jelly. *Figures 3* and *4* show typical results using the protocols above.

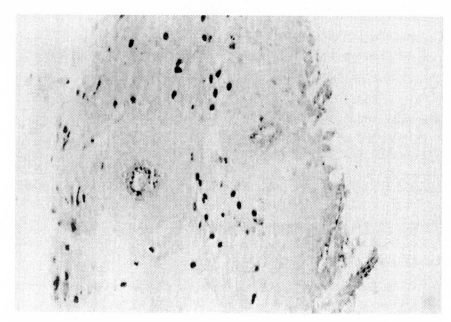

Figure 3. Positive nuclear hybridization signal for human papillomavirus type 6 in a cervical condyloma .X200.

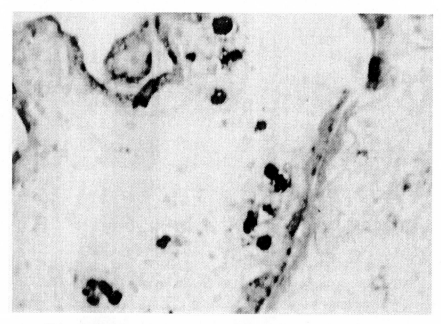

Figure 4. CMV detected by *in situ* hybridization in human placenta .X400.

7. Commercial kits for *in situ* DNA virus hybridization

A number of kits for *in situ* hybridization, localization, and typing of viral infections are now available. The Vira type HPV typing kit from Digene and the HPV *in situ* typing kit from Biohit give good results.

8. Combined *in situ* hybridization and PCR analysis

The polymerase chain reaction (PCR) permits the selective *in vitro* amplification of a particular DNA region by mimicking the phenomena of *in vivo* DNA replication (16). The following reaction components are required: single-stranded DNA template, primers (oligonucleotide sequences complementary to the ends of a defined sequence of DNA template), deoxynucleotide triphosphates (dNTPs), and a DNA polymerase enzyme (*Taq* DNA polymerase). A new DNA strand complementary to the desired template can then be enzymatically synthesized under appropriate conditions. In addition, messenger RNA amplification can be achieved by generation of a cDNA using the enzyme reverse transcriptase (see Chapter 8 this volume).

PCR has been extensively used to amplify specific DNA sequences for use in the molecular analysis of many diseases. PCR can be used to amplify DNA even from samples of formalin-fixed paraffin wax embedded tissue, although amplification failure does occur with this type of material (17). However, one limitation for the histopathologist is that it has not been possible to localize amplified DNA sequences in cells or tissue sections. Recently, however, there have been several studies describing a new technique which combines PCR with *in situ* hybridization permitting the localization of specific amplified DNA segments within isolated cells and sections of tissues. This technique is called *in situ* PCR or PCR *in situ* hybridization.

The technique was first described in 1990 by Haase *et al.* (18) who amplified lentiviral DNA in infected cells and subsequently detected the amplified DNA using *in situ* hybridization. Nuovo *et al.* (19, 20) have further modified and developed the technique for the identification of different types of human papillomavirus in formalin-fixed paraffin wax embedded samples. *In situ* PCR refers to using a biotin or digoxigenin labelled nucleotide directly in the PCR and subsequently detecting the labelled amplified product using standard *in situ* detection protocols (detailed earlier), whilst PCR *in situ* hybridization refers to PCR amplification on a tissue section/cell suspension, followed by application of a biotin or digoxigenin labelled probe to the amplified product and detection of the hybrid by standard *in situ* detection protocols (detailed earlier).

8.1 Theory of *in situ* PCR or PCR *in situ* hybridization

Why should the technique work? Many people question the scientific basis of the *in situ* PCR process, with many centres reporting failure of the technique. In theory, a fixed cell should act like a sponge, or at least a semipermeable dialysis bag. It seems reasonable, therefore, to assume that PCR reagents (*Taq*, primers, etc.) will be able to diffuse through the cell membrane and into the cytosolic and nuclear component of the cell. From provisional results using this technique, it seemed that the majority of the amplified product appeared to localize in the nuclear or cytoplasmic compartments of the cell, although some leakage did occur into the surrounding medium. Hybridization of the amplified DNA/cDNA with a suitably labelled probe was then required immediately, before the amplified product diffused from its site of amplification. The technique can be conveniently performed on a normal, conventional, thermal cycling block or in specially designed thermal cyclers for slide PCR or in a microprocessor controlled oven. This technique is described by Nuovo (21) as 'PCR *in situ* hybridization'.

8.2 Problems associated with the reaction

We evaluated the use of PCR *in situ* hybridization with paraffin wax embedded cell lines for the detection of low and intermediate copy number HPV infection in virally infected cells. To achieve this, three of the authors (JO'L, MC, IH) have examined the SiHa cell line that contains 1–2 copies of HPV 16 and the CaSki cell line which contains approximately 200–300 copies of HPV 16. The 1–2 copies of HPV 16 in SiHa cells cannot be routinely detected by normal non-isotopic *in situ* hybridization (NISH) protocols except for complex five-step procedures (*Protocol 13*). As can be seen in *Figures 5a, b* and 6, using *in situ* PCR, signals can be detected both in SiHa and CaSki cells. However, problems of localization of signal in CaSki cells are noted (22).

The following points should be considered when using PCR *in situ* hybridization or *in situ* PCR:

(a) Amplification of target DNA associated with denaturation steps must not destroy cellular morphology, so as to prevent histological examination of the cells after detection of the amplicons by *in situ* hybridization.

(b) For all reagents (Mg^{2+}, *Taq* polymerase, primers, etc.) the optimum concentrations must be exactly determined for each particular amplification reaction.

(c) The success of *in situ* PCR depends on the amplified DNA remaining localized. It is not clear how amplified DNA would not diffuse away from the site of amplification, although it is probable that networks of amplified DNA form, resulting in the development of an insoluble high molecular weight DNA complex. It has been hypothesized that amplified large DNA fragments would be less likely to diffuse than smaller fragments.

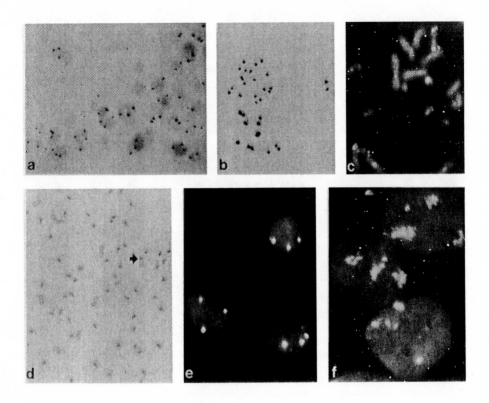

Plate 1 (see Chapter 1 for protocols and references). *In situ* hybridization (ISH) using different labels for non-radioactive probes (see *Table 2*) and different detection systems (see *Table 3* and *5*). (a–d) Detection based on peroxidase and alkaline phosphatase. (e–l) Fluorescence *in situ* hybridization (FISH) based on fluorochrome labelled antibodies or affinity molecules. (a) ISH on a paraffin-embedded tissue section with a biotinylated probe for chromosome 1 and detection using peroxidase with H_2O_2/DAB as substrate (59). (b) Hybridization on a tumour cell line T24 with a chromosome 1 probe (biotinylated) and detection with BCIP/NBT as substrate for alkaline phosphatase. (c) Cosmid hybridizations (two probes both biotinylated) on human chromosome 11 pter and qter. Detection with naphthol/Fast Red as fluorescent label for alkaline phosphatase (60). (d) Double-target hybridization on human/mouse cell line with biotinylated mouse satellite DNA probe and chemically modified total human DNA probe (mercuration). Double-target detection by combined enzymatic detection using peroxidase H_2O_2/DAB (human sequences, brown colour) and alkaline phosphatase BCIP/NBT (mouse sequences, blue colour). Arrow indicates translocation between mouse and human sequences (54). (e) FISH showing trisomy for chromosome 7 in a bladder carcinoma (biotin labelled probe, DNA counter staining PI. (f) Double-target FISH on isolated cells from solid tumour showing imbalance between copy numbers for chromosomes 1 and 18. Probes were labelled with different labels and detected with FITC and TRITC. DNA counter staining with DAPI (61).

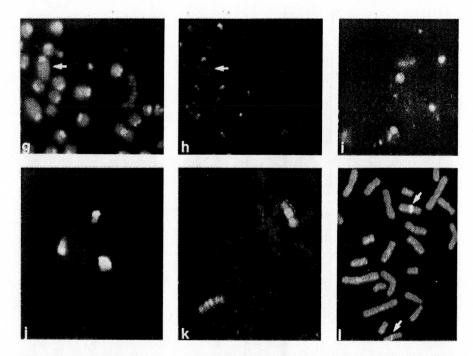

Plate 1 (cont) (g,h) Double-target hybridization with same probes and cell line as (d) (61) with biotinylated mouse satellite DNA probe and chemically modified total human DNA probe (mercuration). Arrows indicate translocation between human and mouse sequences. (g) DAPI general DNA staining of chromosomes (h) FISH results on same field as (g), green fluorescence human chromosomes, red/orange fluorescence mouse sequences. (i) Tripe-target FISH with three different labelled chromosome specific probes on interphase nuclei. Detection with FITC, TRITC, and AMCA fluorochromes (55). (j) Illustration of co-localization of two different probes indicating a translocation between different chromosomes (62). (k) Hybridization with an Alu-PCR labelled human/mouse hybrid cell line containing human chromosome 7 on human metaphase chromosomes. (l) Direct labelling of target sequences by the PRINS reaction (see *Protocol 4*) using an oligonucleotide specific for the repetitive sequences on chromosome 9 (heterochromatin 9q, see arrows) and FITC-dUTP as substrate for the DNA polymerase reaction.

Plate 2 (see Chapter 2 for protocols). Examples of fluorescent *in situ* hybridization. (a) Metaphase spread from a lymphocyte preparation hybridized to an approx. 100 kb cosmid contig specific for chromosome 13. The probe was labelled with FITC, and the metaphases counterstained with propidium iodide. An interphase nucleus showing two corresponding hybridization signals is also present. (b) Aneuploidy detection (XXY). Cultured amniocytes were hybridized with an X cosmid probe labelled with FITC, and a Y cosmid probe labelled with Cy3 (DAPI counterstain). Both interphase nuclei and a metaphase spread clearly show 2 X hybridization signals, and 1 Y hybridization signal. (c) Aneuploidy detection (XXX). Uncultured amniotic fluid cells hybridized to an X cosmid probe labelled with FITC show three hybridization signals in the interphase nucleus. No counterstain was used. (d) Translocation identification. In this example a metaphase spread was hybridized to a FITC-labelled chromosome 15 paint, and a Cy3-labelled Y specific cosmid. (e) High resolution mapping. DNA molecules (each approx. 35 kb) were extended by histone depletion, and hybridized to three members of a chromosome 18

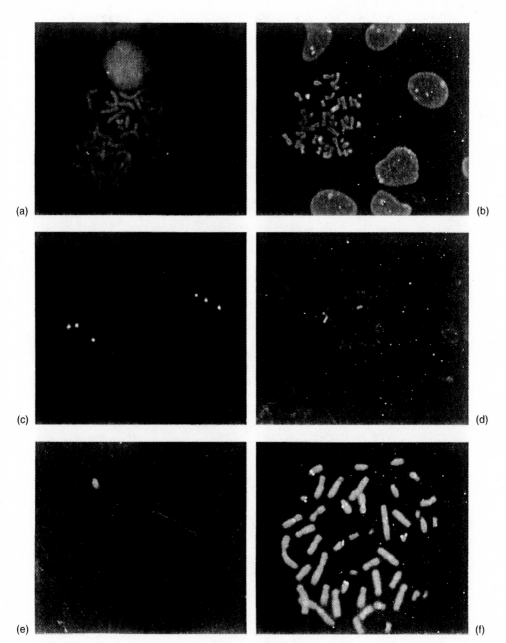

cosmid contig. The two external members were labelled with FITC, while the internal cosmid was labelled with Cy3. Suppression of the repetitive elements yields a characteristic 'beads on a string' appearance, while a large repetitive stretch of DNA is apparent (by the absence of signal) in the middle of the red labelled cosmid. Note that the cosmid order is ambiguously obtained. (f) Simultaneous multicolour hybridization. Cosmids specific for chromosome 13 (pseudocoloured white), 18 (pseudocoloured pink), 21 (pseudocoloured green), and X (pseudocoloured orange) were hybridized to a normal female metaphase.

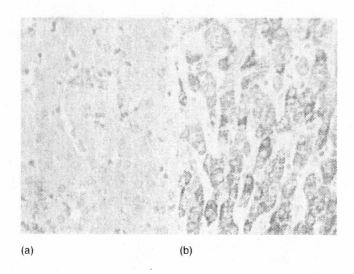

(a)   (b)

Plate 3 (see Chapter 4 for protocols). Demonstration of endogenous biotin in formalin-fixed paraffin sections of human liver. *Plate 3a* shows the signal obtained using a strepta-vidin and biotinylated alkaline phosphatase detection system following proteinase K pre-treatment and overnight hybridization without probe. In contrast, *Plate 3b* shows the level of background obtained with an anti-fluorescein detection system also visualized with alkaline phosphatase using an indirect method. The alkaline phosphatase signal was developed using Napthol phosphate and Fast Red TR substrates and the tissues were counterstained using weak haematoxylin staining.

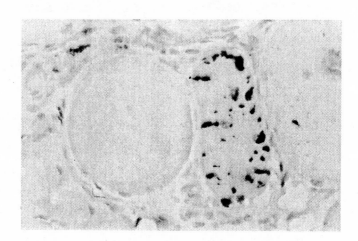

Plate 4 (see Chapter 4 for protocols). *In situ* hybridization following RNase pre-treatment of sections with oligonucleotides to coxsackie virus RNA on paraffin sections of inclusion body myositis. The alkaline phosphatase staining with BCIP and NBT shows discrete signal within muscle fibres, particularly near to vacuoles and within nuclei. This signal is an artefact caused by non-specific binding of the oligonucleotide probes to inclusion bodies and could not be removed by pre-treatments.

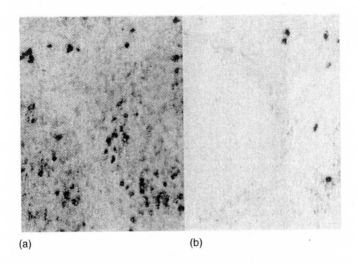

(a)   (b)

Plate 5 (see Chapter 4 for protocols). Low power photomicrographs showing *in situ* hybridization to human immunoglobulin light chain mRNA on a paraffin-embedded lymphoid tissue biopsy taken from a patient with suspected B-cell lymphoma. *Plate 5a* shows kappa mRNA staining in the majority of cells in the germinal centre and extra follicular areas. *Plate 5b* shows comparative lambda mRNA signal in the mantle zone lymphocytes and occasional extra follicular plasma cells. The results confirm the diagnosis of lymphoma and demonstrate light chain restriction under circumstances where immunochemistry would be technically difficult because of the extracellular immunoglobulin.

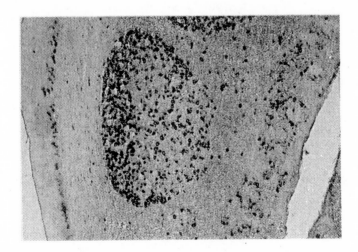

Plate 6 (see Chapter 4 for protocols). *In situ* hybridization to histone mRNA on lymphoid tissue showing proliferation in germinal centres. The cells labelled for histone mRNA are in 'S phase' of the cell cycle.

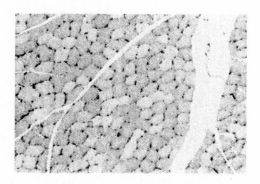

Plate 7 (see Chapter 4 for protocols). *In situ* hybridization of formalin-fixed sections of rat skeletal muscle with oligonucleotides to glyceraldehyde-3-phosphate dehydrogenase mRNA. The pattern of staining reflects the fibre type of the muscle, indicating higher levels of expression in the fast fibres while lower levels are recorded in the slow fibres.

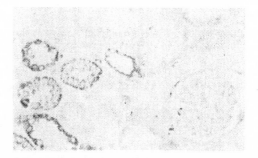

Plate 8 (see Chapter 4 for protocols). *In situ* hybridization detection of the rRNAs within the mitochondria in a paraffin section of human kidney. The alkaline phosphatase signal shows that the distal tubules have a higher density of mitochondria than the proximal tubules. This probe can be used to confirm RNA preservation in the tissue as well as the organelle distribution.

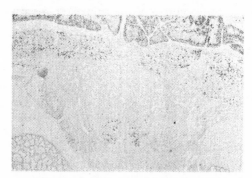

Plate 9 (see Chapter 4 for protocols). Detection of coxsackie virus, using strain-specific oligonucleotides, showing the presence of virus RNA in the anterior horn regions of the spinal cord in a mouse model system. This model system allowed the development of probes to various picorna viruses confirming the specificity and tissue trophisms.

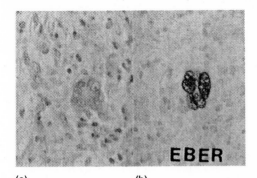

Plate 10 (see Chapter 4 for protocols). Detection of Epstein–Barr virus in Hodgkin's disease using oligonucleotides to EBER nuclear RNAs on paraffin-embedded tissue sections. *Plate 10a* shows the haematoxylin–eosin stained tissue with a characteristic Reed–Sternberg cell. *Plate 10b* shows the in situ signal for EBER in the nucleus of this diagnostic cell.

(a)   (b)

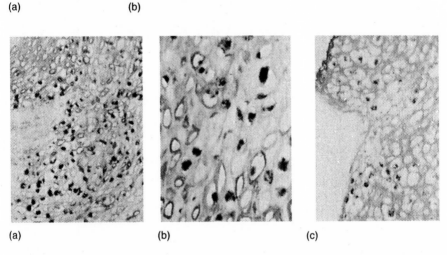

(a)   (b)   (c)

Plate 11 (see Chapter 5 for protocols). Condyloma infected with human papilloma virus (HPV). Frozen section, acetone fixation. (a) Co-localization of cytokeratin 10 and HPV 6 DNA with IPO (DAB) (*Protocol 9*) and DISH (NBT/BCIP) (*Protocols 16* and *17*) respectively. (b) Combination of IAP for cytokeratin 10 (red) (*Protocols 13* and *14*) and DISH (NBT/BCIP) (*Protocols 16* and *17*) for HPV 6 DNA. (c) Combination of IBG for keratin staining (blue-green) (*Protocol 15*) with DISH (NBT/BCIP) (*Protocols 16* and *17*) for HPV 6 DNA.

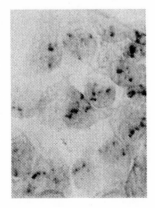

Plate 12 (see Chapter 5 for protocols). CaSki cells, containing HPV 16 DNA, grown on coverslips and fixed in acetone. Co-localization of nuclear HPV 16 DNA (NBT/BCIP) and cytoplasmic cytokeratin (IPO, DAB) (*Protocols 9* with *16* and *17*).

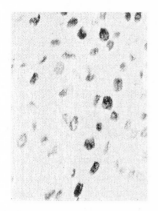

Plate 13 (see Chapter 5 for protocols). Paraffin section of HPV 11-infected tissue, formalin fixation. Combination of DISH (DAB/Ni) for HPV 11 DNA and IPO (DAB) for the viral capsid antigen, both with nuclear localization. Not all cells express the antigen (combination of *Protocols 10* with *18* and *19*).

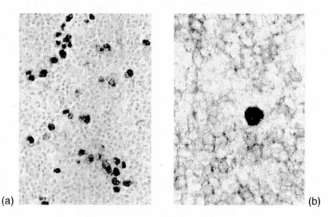

(a)   (b)

Plate 14 (see Chapter 5 for protocols). Formalin-fixed, paraffin-embedded section of Hodgkin's lymphoma. (a) Reed–Sternberg cells express nuclear Epstein–Barr virus (EBER) RNA (black, DAB/Ni with Ag) and cytoplasmic EBV latent membrane protein (RISH *Protocol 23* with *19* and *20* combined with *Protocols 11* and *12*). (b) Reed–Sternberg cells express nuclear Epstein–Barr virus (EBER) RNA (NBT/BCIP). Surrounding cells express cell surface antigen specific for B lymphocytes (monoclonal antibody L26, DAB) (*Protocols 22* with *17* in combination with *11* and *12*).

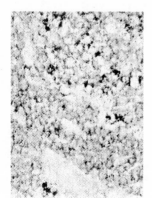

Plate 15 (see Chapter 5 for protocols). Formalin-fixed, paraffin-embedded lymph node of sarcoidosis patient. Expression of beta-2-microglobulin mRNA (black) with B lymphocyte-specific antigen (brown). Some B as well as non-B lymphocytes show high expression of the mRNA (RISH *Protocol 23* with *19* and *20* combined with *Protocols 11* and *12*).

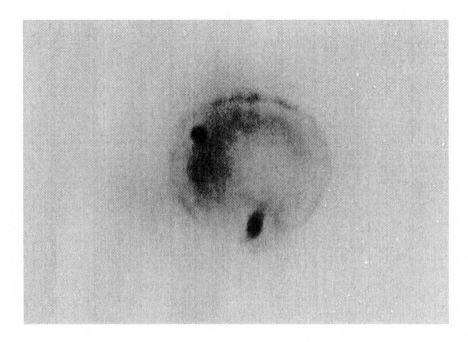

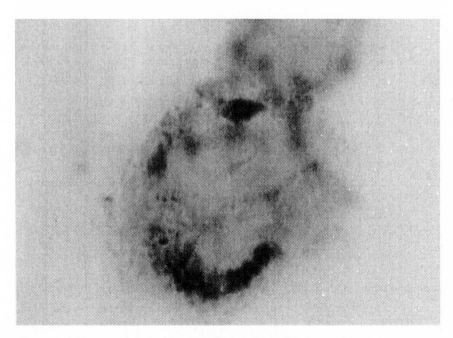

Figures 5a and b. PCR *in situ* hybridization in SiHa cells showing two copies of HPV 16/SiHa cell nucleus. Reproduced with permission (22)

Figure 6. PCR *in situ* hybridization in CaSki cells for detection of HPV 16, illustrating the 'over-spill' phenomenon seen when amplifying high copy numbers of target sequences by this method. Reproduced with permission (22)

However, some authors have suggested that the length of DNA amplified fragment does not affect the final localization of DNA product after *in situ* hybridization.

(d) Tissue drying and loss of tissue adherence during amplification procedures must be prevented.

(e) Failure of the technique. Many centres report consistent failure of the technique. It is possible that the bulk of reagents used in *in situ* PCR protocols may become sequestered during the actual procedure itself. Reagents (especially *Taq* and oligonucleotide primers) may be sequestered by any one of the following mechanisms:

   i. glass

   ii. reactive groups due to silanization of slides

   iii. reactive groups from fixation processes.

(f) Patchy amplification. Even after rigorous precautions during set-up, patchy amplification in some cells is seen. Amplification appears to occur preferentially in some parts of the tissue section/cell suspension. There are several possible reasons for this, including inadequate digestion during the pre-treatment protocol, or lingering DNA cross-linking phenomena because of prior fixation of cells. In addition, the concept of the cell as a sphere must be addressed when one considers that some of the cells have been cut by microtome blades and are now truncated spheres. This would immediately allow free drainage of amplified product from the cell. In addition, the thermal profile of the reaction may differ at several sites on the tissue section/cell suspension.

(g) Diffuse signals should be interpreted carefully. Non-specific binding of primers and detection reagents can occur and thus must be allowed for.

(h) The inclusion of appropriate positive and negative controls: it is extremely important to include all appropriate controls, both positive and negative, and ideally electrophoresis should be carried out to demonstrate that the amplified segment of DNA is of the appropriate molecular weight. In addition, omitting PCR should result in reduced or absent hybridization signals as compared with the *in situ* PCR result.

8.3 Preparation of glass slides for *in situ* PCR/PCR *in situ*

Paraffin embedded tissues or cell suspensions should be placed on a slide pre-treated with organosilane (*Protocol 1*).

8.4 Fixation of cells/tissues and proteolytic digestion

From the work of Nuovo *et al.* (21), it appears that fixation of cells or tissues in buffered formalin for 24 hours and digestion with trypsinogen or proteinase K for 10–15 min is adequate.

8.5 Amplifying solution

The concentration of some of the reagents in the amplifying solution for PCR/*in situ* hybridization/*in situ* PCR vary compared with those for standard PCR. The optimum concentrations as found by Nuovo are detailed in *Table 3*. It is important to note that the optimum concentrations of magnesium and *Taq* polymerase are greater than those for standard PCR. This reflects the difficulty of entry of these reagents to the site of DNA amplification and sequestration of magnesium by cellular components and/or non-specific binding of *Taq* polymerase to the glass slide.

8.6 Primers

Much discussion has taken place, concerning the use of single or multiple primer pairs for the performance of PCR *in situ* hybridization/*in situ* PCR. Haase *et al.* (18) described the procedure of PCR amplification in cell suspensions using multiple primers which dictated the synthesis of overlapping fragments that could form a complex of over 1500 bp. It was proposed that this fragment could not pass through the nuclear membrane, whereas fragments smaller than 450 bp were membrane permeable. This apparent need

Table 3. Optimum concentration of amplifying solution

Magnesium	4.5 mM
Buffer [a]	10 mM Tris–HCl, pH 8.3, 50 mM KCl, 0.001% gelatin
dNTP [b]	200 μM each
Primers	1–5 μM
Taq polymerase	2.5 units/12.5–25 μl

[a] Buffer II, Perkin-Elmer Cetus.
[b] The inclusion of a labelled nucleotide in the amplifying solution (biotin *or* digoxigenin oligonucleotide) may also be used. This is to avoid application of probe post-amplification, and thereby simplifies the post-amplification detection procedure. (This is called '*in situ* PCR' by Nuovo (21).)

for multiple primer pairs limited the applicability of the PCR *in situ* technique because of the high cost and because of the difficulty of generating multiple sequence specific primers for highly polymorphic targets. However, one does not need to use multiple primer pairs to obtain success in PCR *in situ* hybridization. A hybridization signal can be generated if one uses the 'hot start' modification of PCR and a single primer pair. This signal is usually stronger than *in situ* PCRs performed using multiple primer pairs and standard PCR *in situ* hybridization.

8.7 *In situ* PCR/PCR *in situ* hybridization techniques

Protocols 16, 17, and *18* describe three methods which may be used. Appropriate positive and negative controls must be included with each run.

Protocol 16. PCR *in situ* hybridization – for tissue sections, cytospins, or cellular smears

Reagents

- xylene
- 95% and 100% ethanol
- trypsinogen (2 mg/ml) in 10 mM HCl
- 100 μM Tris–HCl, pH 7.0, 100 μM NaCl
- PCR buffer (*Table 3*)

- dNTP solution: stock 10 mM each
- MgCl$_2$: stock solution 25 mM
- primers: stock solution 20 μM each
- *Taq* polymerase
- mineral oil

Method

1. Place 4 μm tissue sections, cytospins, or cellular smears on silane-coated glass slides. Cut a piece of aluminium foil to slightly larger porportions than the slide. Using an immunohistochemical marking pen (DAKO) circumscribe the area occupied by the tissue section/cell suspension.

---

2. For cytospins and cellular smears, fix in buffered formalin overnight.

3. Remove paraffin wax from paraffin wax embedded tissues by washing the slides in xylene for 5 min, followed by washing the slides in 100% ethanol for 5 min, 95% ethanol for 5 min and then air-drying.

4. Treat the slides with trypsinogen (2 mg/ml) in 10 mM HCl for 10–15 min.

5. Inactivate trypsinogen in 100 μM Tris–HCl, pH 7.0, 100 μM NaCl solution.

6. Wash for 1 min in 95% ethanol and 1 min in 100% ethanol and air-dry the slides.

7. Prepare the following amplifying solutions:
   In tube 1:
   - buffer II [a]   2.5 μl
   - $MgCl_2$ 4.5 μl
   - dNTP   4 μl
   - primer 1   1 μl
   - primer 2   1 μl
   - sterile water   11 μl

   Remove 5 μl and place into a second tube, tube 2. Keep this tube in ice. (This is enough for two reactions.)

8. Divide the remaining 19 μl equally between two sections.

9. Overlay with a plastic coverslip and anchor to the slide at one edge with nail polish.

10. Place the slides (in aluminium foil boats) on the heating block of the thermal cycler at 82°C for 7 min.

11. Add 1 μl of *Taq* DNA polymerase to tube 2. Keep on ice.

12. Once the slide has reached 80°C, lift one edge of the coverslip gently and add 2.5 μl from tube 2 per section.

13. Overlay with pre-heated mineral oil at 82°C.

14. Perform the following cycling parameters:
    1 cycle of:
    - 94°C for 3 min
    20–40 cycles:
    - 94°C for 1 min
    - 55°C for 2 min

15. At the end of cycling, remove the coverslip and nail polish. The coverslip can be removed using a scalpel or acetone.

16. Remove the mineral oil by placing the slides in xylene for 2 min followed by a 2 min wash in 100% ethanol.

Protocol 16. *Continued*

17. Allow the slides to air-dry.

18. Proceed as in *Protocols 6* and *7* (as described earlier) for the hybridization of biotin *or* digoxigenin labelled probe to amplified DNA.

19. Detect the hybrid using *Protocols 8, 9, 12,* or *13* as appropriate. Shorter detection protocols are favoured.

ᵃ Buffer II, Perkin-Elmer Cetus

Inclusion of a labelled nucleotide in the amplifying solution is a major modification of the technique described above. Digoxigenin-11-dUTP may be incorporated into the amplified product during standard PCR.

Protocol 17. *In situ* PCR

Reagents

- see *Protocol 16*
- Digoxigenin-11-dUTP (Boehringer–Mannheim)
- 2% BSA, 0.1% SSC
- anti-digoxigenin conjugated to alkaline phosphatase (Boehringer–Mannheim)
- TBS (*Table 2*)
- TBS + 50 μM MgCl$_2$
- NBT/BCIP detection reagents (*Table 2*)

Method

1. Proceed as in *Protocol 16* (Steps 1–14) except in the amplifying solution, add 10 μM of DIG–dUTP to the amplifying mixture. Note that this is a ratio of 1:20 compared to dTTP.

2. Following completion of the amplifying reaction and removal of coverslip, wash the slides for 10 min at 45°C in a solution of 2% BSA and 0.1 × SSC to remove unincorporated DIG–dUTP.

3. Place the slides in a moist chamber and incubate in anti-digoxigenin–alkaline phosphatase conjugated antibody, 1/50 dilution in TBS at room temperature for 30 min.

4. Incubate the slides in TBS and 50 μM MgCl$_2$ at room temperature for 2 min.

5. Place the slides in the development reagent for NBT/BCIP at room temperature and monitor colour development for 15 min–2 hours.

Tissue preparation and the pre-treatment of paraffin wax embedded cell lines is the same as for the pre-treatment of fixed paraffin-embedded tissues as described in *Protocol 5*. Alternatively, this procedure can be simplified by following the de-waxing steps in the above protocol and proceeding directly to proteinase K digestion.

Protocol 18. PCR *in situ* hybridization on paraffin wax embedded cell lines

Reagents

- PCR solution: 50 mM KCl, 10 mM Tris–HCl, pH 8.3, 4–5 mM MgCl$_2$, 0.01% gelatin, 200 μM of each dNTP, 5 μM primers.
- *Taq* polymerase
- mineral oil
- ethanol
- 2% paraformaldehyde

Method

1. Place the slide containing the fixed cell suspension in an aluminium foil boat, trimmed to slightly larger proportions than the slide.

2. Add 10 μl of the PCR solution on top of the cell suspension contained in the deep well slide. Cover the well with a pre-cut piece of gel bond, hydrophobic side down.

3. Place the slide (in the aluminium foil boat) on the heating block of the thermal cycler and allow the temperature to increase to 80°C.

4. Once the temperature has reached 80°C, lift a corner of the gel bond and add 2.5 μl *Taq* DNA polymerase (2.5 units per 12.5 μl) to the PCR mix contained on the slide.

5. Replace the coverslip and seal the margins with nail polish. Add 1–2ml of pre-heated mineral oil (80°C) on top of the slide to ensure optimum thermal kinetics. Apply the following PCR protocol:
   - 94°C for 6 min
   - 40 cycles of 55°C for 2 min
   - 94°C for 1 min.

6. Following amplification, dip the slide in chloroform to remove the mineral oil and carefully remove the gel bond coverslip. Gel bond, hydrophobic side towards the cell suspension, allows the PCR reagent mix to remain on the tissue in the well.

7. Carefully dip the slides in 100% alcohol and dehydrate. The slides can also be post-fixed in 2% paraformaldehyde to maintain localization of PCR product.

8. Proceed as in *Protocols 6* and *7* for hybridization of biotin or digoxigenin labelled probe to amplified DNA.

9. Detect the hybrid using *Protocols 8, 9, 12,* or *13* as appropriate. Shorter detection protocols are favoured.

8.8 Helpful hints when performing PCR *in situ* hybridization/*in situ* PCR

(a) The precise volume of PCR reactants to be used is probably determined by the surface area of the cytological/biopsy material that is to be examined. For large biopsies/cervical biopsies, 50 μl of PCR reactants are required. Cytological smears and small cell suspensions probably only require between 10 and 15 μl.

(b) The leakage of amplification product (as described in CaSki cells) depends not only on the amplicon size but also on the degree of porosity of the nuclear and cytoplasmic membranes, following proteolytic digestion of the cell to allow access of PCR reagents to the nucleic acid template before proceeding with PCR. In order for the amplified product to remain localized within the nucleus, electrostatic forces or DNA cross-linking phenomena must be induced. In our experience (J. O'Leary, in preparation), products of the order of 100 bp tend to diffuse from the nuclear to the cytoplasmic compartment of the cell quite readily following amplification. Improvement can be achieved however, by post-fixing the slides containing the amplified product with 2% paraformaldehyde. In addition, a dehydration step of 100% alcohol may also help in the localization of the product. Paraformaldehyde induces cross-linking of newly formed nucleic acid template to histone-protein subfractions and to the original nucleic acid template present in the nuclear compartment of the cell. Larger fragment amplicons (300–500 bp) appear to remain within the nuclear compartment even in the absence of post-fixation.

(c) The use of a flexible coverslip on the PCR reactants during PCR *in situ* hybridization/*in situ* PCR is also advised. Gel bond (FMC Bioproducts), used in the *in situ* hybridization protocols detailed earlier, has a hydrophilic and hydrophobic side and is quite useful in this regard. It is advisable to place the hydrophobic side down, so that when the coverslip is lifted to place primers and *Taq* DNA polymerase on the section during the 'hot start' procedure, no PCR reactants are lost or adhere inadvertently to the coverslip.

(d) Ideally *in situ* PCR should be performed in deep welled slides, as for conventional *in situ* hybridization analysis. The concept of heat transfer between the metal block and the glass slide must be addressed. To achieve this and to maximize the rate of heat transfer, the glass slide (containing the tissue section) is placed in an aluminium foil boat which is then placed in the heating block of the thermal cycler. To optimize heat transfer kinetics the slide is then overlaid with pre-heated mineral oil at 80°C. The temperature of the slide should ideally be examined using a thermocouple.

Currently, many problems exist with the technique and some refinement of technology is required.

Acknowledgements

We wish to thank Juliet Hamblin who typed the script and Steve Toms who undertook the photographic work.

References

1. Ouder, P. and Schatz, C. (1985). In *Nucleic acid hybridization: a practical approach* (ed. B. D. Hames and S. J. Higgins), pp. 161–78. IRL Press, Oxford.
2. Brack, C. (1981). *Crit. Rev. Biochem.*, **10**, 113.
3. Wells, M., Griffiths, S., Lewis, F., and Dixon, M. F. (1987). *J. Pathol.*, **152**, 77.
4. Terry, R. M., Lewis, F. A., Griffiths, S., Wells, M., and Bird, C. C. (1989). *Clin. Otolaryngol.*, **157**, 109.
5. Quiney, R. E., Wells, M., Lewis, F. A., Terry, R. M., Michaels, L., and Croft, C. B. (1989). *J. Clin. Pathol.*, **42**, 694.
6. Rapp, A. K., Geelen, J. L., Van der Meer, J. W. M., Van de Rijhe, F. M., Van den Boogart, P., and Van der Ploeg, M. (1988). *Histochemistry*, **88**, 367.
7. Sequiera, I. W., Jennings, L. C., Carrasco, L. H., *et al.* (1979). *Lancet*, **ii**, 609.
8. Weiss, L. M., Strickler, J. G., Warnke, R. A., Purtils, P. T., and Sklar, J. (1987). *Am. J. Pathol.*, **129**, 86.
9. Murphy, J. K., Young, L. S., Bevan, I. S., Lewis, F. A., Dockey, D., Ironside, J. W., O'Brien, C. J., and Wells, M. (1990). *J. Clin. Pathol.*, **43**, 220.
10. Landers, R. J., O'Leary, J. J., Crowley, M., Bailey-Healy, I., Lewis, F. A., and Doyle, C. T. (1993). *J. Clin. Pathol.*, **46**, 931.
11. Lewis, F. A., Griffiths, S., Dunnicliffe, R., Wells, M., Dudding, N., and Bird, C. C. (1987). *J. Clin. Pathol.*, **40**, 163.
12. Herrington, C. S., Burns, J., Graham, A. K., Bhatt, B., and McGee, J. O'D. (1989). *J. Clin. Pathol.*, **42**, 592.
13. Herrington, C. S., Burns, J., Graham, A. K., Bhatt, B., and McGee, J. O'D. (1989). *J. Clin. Pathol.*, **42**, 601.
14. Herrington, C. S., Graham, A. K., and McGee, J. O'D. (1991). *J. Clin. Pathol.*, **44**, 33.
15. Herrington, C. S., de Angerlis, M., Evans, M. F., Troncone, G., and McGee, J. O'D. (1992). *J. Clin. Pathol.*, **45**, 385.
16. Saiki, R. K., Scharf, S., Falona, F., Mullis, K. B., Horn, G. T., Erlich, H. A., and Arnheim, N. (1985). *Science*, **230**, 1350–4.
17. Doyle, C. T. and O'Leary, J. J. (1992). *J. Pathol.*, **166**, 331.
18. Haase, A. T., Retzel, E. F. and Staskus, K. A. (1990). *Proc. Natl. Acad. Sci. USA*, **87**, 4971–5.
19. Nuovo, G. J., Gallery, F., MacConnell, P., Becker, P., and Bloch, W. (1991). *Am. J. Pathol.*, **1349**, 1239–44.
20. Nuovo, G. J., MacConnell, P., Forde, A., and Delvenne, P. (1991). *Am. J. Pathol.*, **139**, 847–50.
21. Nuovo, G. J. (ed.) (1992). *PCR in-situ hybridisation*. Raven Press, New York.
22. O'Leary, J. J., Browne, G., Johnson, M. I., Landers, R. J., Crowley, M. *et al.* (1994). *J. Clin. Pathol.*, **47**, 433.

Non-isotopic detection of RNA
in situ

JAMES HOWARD PRINGLE

1. Introduction

The synthesis of RNA plays a crucial role in the molecular biology of the cell and is, therefore, an important process to study. Although RNA can provide a source of genetic information in the genome of RNA viruses, its main role in the cell is to aid the transfer of genetic information from genomic DNA to functional proteins during gene expression. The process is regulated at each step in the pathway from DNA to RNA to protein (1). This includes the control of transcription of heterogeneous nuclear RNA (hnRNA), hnRNA processing to produce mature mRNA via the spliceosomes, mRNA transportation to the cytoplasm, and mRNA translation into proteins. The majority of cellular RNAs including tRNAs, rRNAs, and small nuclear RNAs (snRNA) have catalytic functions in this process (2, 3).

In multicellular organisms the majority of expressed genes are common to all cells. Only 2–3% of these genes are differentially expressed and account for cell specialization. As most tissues are composed of a variety of cell types in close proximity, analysis of the differentially expressed genes can only be reliably achieved using *in situ* methods. Once the RNAs have been extracted from the tissue for Northern blotting or reverse transcriptase-polymerase chain reaction (RT-PCR) (ref. 4 and see Chapter 8 of this volume), then information about the site of synthesis is lost. Therefore, knowledge of the spatial distribution and quantification of RNAs at the tissue, cellular, and subcellular levels is needed to investigate gene expression and cell differentiation successfully.

In this chapter, RNA *in situ* hybridization methods that involve non-isotopic detection will be described. The advantage of non-isotopic techniques is that they provide a reliable method for accurate localization of the target RNA sequences which can be analysed quantitatively if necessary. A variety of visualization techniques are available which allow the end user to adopt the best method which suits the target and the biological samples chosen. The cellular distribution of RNA can be studied in relation to physiological, pathological, toxicological, and developmental processes. Examples of these applications will be included to emphasize the variety of information available from studies of RNA *in situ*.

Although RNA is inherently more prone to degradation than DNA there are a number of advantages in choosing this molecule as a target for *in situ* hybridization. RNA is usually more abundant than its homologous DNA sequence and is, therefore, easier to detect. This is true of many DNA viruses where the genome is maintained in low copy number during latency while viral RNAs are transcribed at abundant levels. RNA targets reveal information about cellular activity or lack of activity for a specific gene. The production of hnRNA confirms that gene expression is switched on or up-regulated. Detection of specific mRNAs provides information about the steady-state translation of gene products in cells. Detection of other RNAs within cells, such as rRNA or snRNA, reflects information about the metabolic activity of the cell. For example detection of mitochondrial rRNA can give information about the numbers of these organelles and their distribution in tissues. RNA expression can also be studied in combination with immunocytochemistry to confirm the site of expression of a specific gene product and to identify the phenotype of the cell.

2. Technical considerations

2.1 Choice of RNA targets

Before developing a new RNA target it is essential to obtain preliminary knowledge of the gene sequence in order to produce the appropriate controls, probes, and protocols. If information about the cellular localization is not necessary then other hybridization methods using extracted nucleic acids may be more appropriate. If there is a choice of gene then select a RNA target that is well expressed. The RNA must be a specific target and probes should not cross-hybridize to related sequences in the cell. Another important requirement before implementing a new RNA *in situ* on unknown material is to develop a model or reference system. This may be a cell line or control tissues from animals or human biopsies. Such a model will allow the preliminary characterization of the probes by Northern blotting (4). This will give clues to the cellular copy number of the target sequences and allow the hybridization conditions to be established in order to avoid cross-hybridization with sequences related, but not identical, to the probes.

The establishment of suitable control cells or tissues enables positive and negative samples to be evaluated with each RNA *in situ* experiment. This provides internal controls and also allows the methods to be developed and optimized so that precious test samples can be conserved for critical experiments. Whenever possible the model should reflect the cellular conditions to be investigated. Often methods are developed using cell lines where cells in culture bear little similarity to cells in tissues. In tissues, cellular contacts and the stromal matrix may influence the effectiveness of the technique.

2.2 Requirements for RNA *in situ*
The following are needed:

(a) a cloned source of target sequence

(b) sequence data from the target nucleic acid

(c) information on the expression of the gene and tissue distribution

(d) Northern blotting data on RNA from cell lines or tissues

(e) information on RNA stability/secondary structure, repeat sequences, exon boundaries, untranslated regions

(f) sequence comparisons with other species

(g) evidence for transcription termination or differential splicing

(h) for viral sequences evidence of strain variation/polymorphisms

(i) control model tissues and cells positive and negative for the target RNAs; these can be verified by reverse transcriptase PCR or Northern blotting

2.3 Probes
There are a variety of methods for enzymatic labelling of DNA and RNA probes with non-radioactive nucleotide analogues (refs 5–7 and Chapter 1 of this volume). In addition non-radioactive probes can be prepared using chemical derivation methods (8–12). In this chapter oligonucleotide probes are described for RNA *in situ*. There are a number of advantages of using oligonucleotides for RNA *in situ* (13–15). A cDNA is not needed, only the published sequence is required, and a number of oligonucleotides can be synthesized to form a probe cocktail representing one RNA target. Since the probes are designed directly from the gene sequence it is possible to avoid secondary structure in the target and sequences common to other unrelated genes. The corresponding sense oligonucleotides can be used as negative controls as long as transcription is confined to one strand of the target gene. The probes can be designed to hybridize to alternatively spliced mRNA, introns for hnRNAs, conserved or specific sequences in a gene family, and RNA from viral or other pathogens. Oligonucleotides are easy to label and are used with a simple hybridization protocol. One of the major advantages of oligonucleotide probes is that they are small single-stranded molecules which penetrate cells and tissues on a microscope slide.

2.4 Controls
Hybridization reactions may produce misleading results due to unexpected homologies between the probe and related non-target nucleic acid sequences. In addition, hybridization between G–C rich regions of the probe and unrelated target sequences may give rise to spurious positive results if the

correct stringency is not used. These cross-hybridization reactions can be avoided by the careful choice of probe sequence and the inclusion of control probes to ensure that the correct target sequences are detected.

Pre-hybridization treatments with RNase or DNase are useful controls to confirm the target nucleic acids as DNA or RNA. Care should be taken however, that a reduction of hybridization signal after nuclease treatment is truly due to removal of the target and not due to destruction of the probe by a residual enzyme.

Apart from cross-hybridization with target nucleic acids, probes may bind to cells through other non-specific means: charge interactions between proteins and probes may be one cause. Other probe–protein interactions may occur with DNA binding proteins. These non-specific interactions can be controlled using a negative control probe which does not hybridize to the cellular nucleic acids. In addition to these effects, a variety of histochemical/immunocytochemical techniques may produce spurious positive results. *Plates 3* and *4* show a number of typical technical artefacts that can be encountered with non-radioactive *in situ* methods. These can be controlled by applying the detection system to tissue sections that have been incubated in hybridization buffer without probe.

Control probes should include sense sequence oligonucleotides to the target gene, antisense oligonucleotides to a gene not expressed in the cells or tissue, a random sequence oligonucleotide, and oligo d(A). Oligo d(T) probes can be used to detect polyadenylated RNA giving valuable information about cellular activity and mRNA preservation as well as confirming that the *in situ* hybridization proctocol is optimized (16).

3. RNA *in situ* methods

3.1 Sample collection

It is essential for RNA *in situ* that tissue is taken as rapidly as possible and fixed immediately and completely from the living state. In practice it can be difficult to achieve this ideal with human tissues or pathological material where prolonged ischaemia, anoxia, and tissue damage often take place. In obtaining tissues every attempt should be made to prevent RNA degradation to preserve cells with intact RNA representing the *in vivo* steady-state conditions. Similarly, cells or cell suspensions should be transferred to fixative as quickly as possible without allowing the cells to be compromised under adverse growth conditions. All specimen handling procedures should be carried out using gloved hands and sterile instruments to avoid ribonuclease contamination.

3.2 Fixation

Effective fixation is one of the most important steps towards obtaining satisfactory *in situ* hybridization results for RNA detection. Unfortunately, it can be the most difficult procedure to control as many specimens are fixed using traditional methods designed for histological preservation often with suboptimum conditions for RNA preservation.

Optimum fixation is required to retain target nucleic acids, preserve tissue and cellular morphology, and allow efficient hybridization. DNA target sequences are well preserved with most fixatives and good results can be obtained as long as the tissue is not excessively cross-linked in aldehyde fixatives. For RNA detection tissues fixed with organic solvent fixatives, such as Carnoy's solution, are known to give poor and erratic results, whereas those fixed with paraformaldehyde produce more consistent results.

Initial experiments on tissue culture cell fixation (17) showed that optimum retention of RNA could be achieved with a short 1–15 min incubation in paraformaldehyde, whereas ethanol/acetic acid or Carnoy's fixative showed loss of up to 75% of total cellular RNA. Recent experiments on RNA integrity and preservation in paraffin-embedded livers suggest that paraformaldehyde-fixed tissues gives optimum RNA retention at about 45% of total cellular RNA compared to less than 10% with Carnoy's (18). Glutaraldehyde fixatives should be avoided for *in situ* hybridization for most applications because of the increased cross-linking produced and the excessive signal to noise ratio.

A variety of fixation methods are available for different specimens. Sacrificed animals may be perfused with buffered paraformaldehyde for optimum fixation of specific organs. Biopsies can either be fixed by immersion into formalin fixatives or rapid frozen for cryostat sections. Cultured cells or cell suspensions can be fixed directly in formalin fixatives and embedded in cytoblocks or transferred to slides by cytocentrifuge.

When applying *in situ* methods to tissue or cells fixed with different protocols it is very important to determine that optimum results have been obtained for each specimen. The differences in fixation between specimens may lead to different hybridization results, either due to direct effects on the retention of RNA or by altering the accessibility of sequence. For reviews of fixation conditions and procedures see references 19–21.

3.3 Slides, section adhesion, and preparation of RNase-free solutions, glassware, and plastics

Protocol 1 describes methods for preparing RNase-free solutions, glassware, and plastics. As with all procedures which involve the handling of RNA, care should be taken to avoid nuclease contamination of either reagents or specimens, which would lead to reduction or loss of signal after hybridization. These precautions are particularly important when dealing with low copy RNA as RNase is ubiquitous and heat stable. Gloves should always be worn when carrying out any of the procedures, and solutions should be sterilized by autoclaving. It is also advisable to treat solutions where appropriate with 0.1% diethyl-pyrocarbonate (DEPC) to inactivate RNase. It is important to ensure that the sections adhere to the slide throughout the many stages of the methodology. Section adhesion can be achieved using a range of suitable adhesives, including poly-L-lysine (22) or aminopropyl-triethoxysilane (23) (*Protocol 2*).

Protocol 1. Treatment of solutions, glassware, and plastics to destroy RNase activity

Reagents

- diethylpyrocarbonate (DEPC) (Sigma)
- ultrapure water
- industrial methylated spirits (IMS)
- 3% H_2O_2

A. *Preparation of diethylpyrocarbonate (DEPC)-treated water*

1. Add DEPC to ultrapure water to a final concentration of 0.1% (w/v).

2. Shake well to dissolve and allow to stand overnight in the fume cupboard.

3. Autoclave the following day.

B. *Preparation of DEPC-treated solutions*

1. Prepare the solutions then follow *Protocol 1A* using the resultant solution in place of ultrapure water. It is important to note that Tris is destroyed by DEPC. Solutions containing this reagent should be made up in RNase-free glassware using water that has been treated with DEPC and autoclaved.

C. *Preparation of RNase-free glassware*

1. Incubate the glassware at 200°C overnight.

D. *Preparation of RNase-free plastics*

1. Rinse in 96% IMS.

2. Soak in 3% aqueous hydrogen peroxide for 10 min.

3. Rinse in DEPC-treated water.

4. Dry and protect from dust.

Protocol 2. Coating slides in the adhesive 3-aminopropyl-triethoxysilane

Reagents

- 1% Lipsol
- industrial methylated spirits (IMS)
- 1% solution of 3-aminopropytriethoxy-silane (APES) in dry acetone (Sigma)

Method

1. Immerse the slides in 1% Lipsol for 30 min at room temperature (RT).
2. Wash in running tap water for 30 min at RT.
3. Wash in two changes of ultrapure water for 5 min each.
4. Wash in two changes of 96% IMS for 5 min each.
5. Air-dry in front of a fan for 10 min.[a]
6. Coat the slides in a freshly prepared 1% solution of APES in dry acetone for 5 sec.
7. Rinse twice quickly in dry acetone.
8. Wash twice in ultrapure water.
9. Air-dry at 42°C overnight.
10. Store at room temperature in a dust-free environment.

[a] If slides are supplied ready washed start at this point in the protocol.

3.4 Frozen tissue for *in situ* hybridization

Tissue biopsies can be snap-frozen in liquid nitrogen, pre-cooled *n*-heptane or isopentane. The aim is to freeze the tissue as rapidly as possible and prevent the formation of large ice crystals which can adversely affect morphology. The method employs a dry-ice/*n*-hexane freezing mixture to freeze the tissue in optimal cutting temperature compound (OCT) (14) on a cork mount. This has advantages over other methods as the tissue is easier to trim and section and can be stored and recut many times, just like a wax block, but still gives excellent RNA retention and acceptable morphology (*Protocol 3*). After sectioning, the slides are placed on dry-ice, to freeze-dry the tissue which is then fixed with paraformaldehyde or formalin (*Protocol 4*).

Protocol 3. Tissue for frozen sections

Equipment and reagents

- cork mounts
- OCT mounts (Gurr, BDH)
- liquid nitrogen
- isopentane
- cryotubes (Nunc)

Method

1. Prepare 5 mm × 5 mm × 2 mm cork mounts using a scalpel blade.
2. Spread a thin (5 mm) layer of OCT compound into the top of each cork mount.
3. Using a liquid nitrogen Dewar pre-cool a beaker containing isopentane until the isopentane begins to freeze (− 150°C).

Protocol 3. *Continued*

4. Using freshly excised biopsies trim the tissue to 3 mm cubes and orientate on the cork mount.

5. Using artery forceps quickly lower the mount into the beaker until the tissue is submerged below the surface of the cold isopentane. Leave it there until the OCT compound is frozen and turns white.

6. Transfer the frozen cork mount with tissue to labelled Nunc cryotubes and store them in liquid nitrogen vapour.

7. Tissue frozen this way can be kept for RNA *in situ* for six months to one year.

Protocol 4. Cryostat sections for RNA *in situ*

Equipment and reagents

- 4% paraformaldehyde in PBS: dissolve paraformaldehyde (Hopkins and Williams) to a concentration of 4% in 50 ml of ultra-pure water with agitation at 65 °C, until the solution remains slightly cloudy. Add to the cloudy solution 2 μl of 5 M sodium hydroxide, cool, then filter through Whatman filter paper. Adjust volume to 100 ml and pH to 7.2–7.4. Store at 4 °C for up to 24 h.

- 10 × phosphate buffered saline (PBS): 1.3 M sodium chloride; 70 mM sodium dihydrogen orthophosphate; 30 mM di-sodium hydrogen orthophosphate, dihydrate
- Nunc cryotubes
- OCT compound (see *Protocol 3*)
- APES-coated slides (see *Protocol 2*)
- alcohol series

Method

1. Wearing goggles and low temperature gloves transfer samples from the liquid nitrogen fridge to a Dewar containing liquid nitrogen.

2. Remove the cork-mounted tissue sample from a Nunc cryotube and transfer it to a cryostat chuck covered with OCT compound. Freeze the specimen in place by immersing the chuck and holder into liquid nitrogen (24).

3. Cut sections for RNA *in situ* at 7–12 μm and adhere to treated slides (24).

4. Freeze-dry the sections on the surface of dry-ice for 20 min.

5. Fix the sections in 4% paraformaldehyde at 4 °C for 10–15 min.

6. Wash the sections in two changes of PBS for 5 min each at RT.

7. Dehydrate through an alcohol series (50%, 70%, then 100%).

8. Slides can be stored in 70% ethanol at 4 °C for several weeks.

9. Remove the cork-mounted tissue sample from the chuck and re-freeze in a Nunc cryotube.

3.5 Wax-embedded cells or tissue for *in situ* hybridization

Immersion fixation, paraffin embedding, and section cutting of tissue biopsies are described in *Protocol 5*. Providing the above conditions are met no special precautions are necessary during subsequent processing as fixed tissue is quite resistant to the effects of RNases. Embedded blocks are very stable and can be stored in cool dry conditions for many years. *Protocol 6* describes the preparation of cytocentrifuged cells and cytoblocks for use in RNA *in situ*.

Protocol 5. Immersion fixation and embedding of tissue biopsies

Equipment and reagents

- 4% paraformaldehyde (see *Protocol 4*)
- 10% formal saline
- Tissue-Tek cassettes
- paraffin
- industrial methylated spirits (IMS)
- xylene

Method

1. Obtain tissue biopsies from animals or humans as rapidly as possible after excision. Where this is not possible transport tissues in transfer media so that the tissues do not dry or suffer heating effects from light.

2. Slice tissues thinly, to a few millimetres, to allow penetration of the fixatives into the tissue and to prevent RNA degradation successfully.

3. Ensure that the tissue is suspended in a large volume of fixative (50–100 ml); cotton wool can be placed in the bottom of the container to allow the fixative to penetrate the specimen uniformly.

4. Fix for 24–48 h in 4% paraformaldehyde or 10% formal saline at ambient temperature. For critical experiments with low copy number RNA targets, fix tissues at 4°C for 12 h followed by fixation at ambient temperature for 36 h.[a]

5. Transfer fixed specimens to Tissue-Tek cassettes. Orientate the tissue appropriately for section cutting and firmly attach the metal lid to the cassette.

6. Dehydrate the tissue and infiltrate with paraffin using an automatic tissue processor (for example Hypercentre from Shandon). The following steps are required:
   - 2 changes of 75% IMS for 1 h, each
   - 2 changes of 95% IMS for 1 h, each
   - 3 changes of 99% IMS for 1 h, each
   - 2 changes of xylene for 1 h, each
   - 2 changes of molten paraffin wax (Paraplast; Sherwood, Medical or equivalent) for 1 h at 60°C

Protocol 5. *Continued*

7. Complete the embedding using an embedding machine with wax baths at 60–65°C and cold plates at 4°C.

8. Cut 4–5 μm sections with a rotary microtome with disposable blades (Feather). Clean the blades with xylene between blocks.

9. Float the sections on warm sterile water (45°C) in a histological water-bath. The water should be changed on a regular basis to avoid contamination with microorganisms.

10. Pick up the sections on aminopropyl-triethoxysilane treated slides (see *Protocol 2*).

11. Dry the slides and sections at 37°C overnight.

12. Store the slides at ambient temperature or 4°C, desiccated. RNA signals are retained on sections stored under these conditions. However antigens may be more labile.

[a] Tissues obtained after perfusion fixation are given a shorter period of immersion fixation at ambient temperatures (12–24 h) before dehydration and embedding.

Protocol 6. Preparation of cytocentrifuged cells and cytoblocks

Equipment and reagents

- buffered trypsin–EDTA
- PBS (see *Protocol 4*)
- APES-treated slides (see *Protocol 3*)
- 4% paraformaldehyde (see *Protocol 4*)
- 10% formal saline
- industrial methylated spirits (IMS)
- Tris buffered saline (TBS): 50 mM Tris–HCl, pH 7.4, 100 mM NaCl
- 70% ethanol
- Shandon cytoblocks

A. *Preparation of cytocentrifuged cells*

1. Release monolayered cells from the surface of the culture flask using buffered trypsin/EDTA or a rubber policeman. Suspension cultures are harvested by centrifugation at 80 g for 5 min.

2. Wash the cells in fresh medium by centrifugation at 80 g for 5 min, resuspend in fresh medium, repellet the cells, and resuspend in PBS.

3. Cytocentrifuge the cells on to aminopropyl-triethoxysilane treated slides (see *Protocol 2*) using the Shandon Cytospin. Load 2×10^5 cells and centrifuge at 800 r.p.m. (30 g) for 10–15 min.

4. Fix the cells in 4% paraformaldehyde or 10% formal saline per well at 4°C for 10–20 min.

5. Wash in PBS for 5 min at 4°C.

6. Dehydrate the slides successively in:
   - 2 changes of 75% IMS for 5 min each
   - 2 changes of 95% IMS for 5 min each
   - 2 changes of 99% IMS for 5 min each
7. Air-dry ready for pre-hybridization or store the slides in 70% ethanol at 4°C. Stored slides are stable for several weeks.

B. *Preparation of cytoblocks*

1. Release monolayered cells from the surface of the culture flask using buffered trypsin/EDTA or a rubber policeman. Suspension cultures are harvested by centrifugation at 80 *g* for 5 min.
2. Wash the cells in fresh media by centrifugation at 80 *g* for 5 min followed by resuspension in fresh medium and then fix in 10% formal saline at ambient temperature for 30 min.
3. Wash the cells twice in TBS by centrifugation at 80 *g* for 5 min.
4. Prepare cytoblocks using the Shandon Cytoblock System according to the manufacturer's instructions. Load 1×10^6 of fixed cells in approximately 20 μl.
5. Centrifuge cells into the cassettes at 60 *g* for 5 min at low acceleration.
6. Leave the cassettes in 70% ethanol or 10% formal saline at ambient temperature overnight prior to embedding.
7. Embed and cut sections as described in *Protocol 5*.

3.6 Pre-treatments

A number of procedures are employed to treat sections after fixation to increase accessibility of the probe and to reduce non-specific binding during hybridization. As with all procedures which involve the handling of RNA, care should be taken to avoid nuclease contamination of either reagents or specimens, which would lead to a reduction or loss of signal after hybridization. Fixation with cross-linking fixatives particularly in wax-embedded tissues retains the RNA but reduces the access of the probe. The aim of pre-treatment is to render the target nucleic acid accessible to the probe. This is done by controlled digestion with a proteolytic enzyme. The use of proteinase K has an advantage over other proteolytic enzymes as during incubation it digests nucleases that might be present in the tissues, it is not autolytic thus giving more reproducible digests, and it is a stable enzyme which can easily be prepared free of nucleases. The amount of proteolysis required will depend on the degree of cross-linking. This will vary from mild digestion for paraformaldehyde-fixed blocks to high concentrations of proteinase K for formalin-fixed pathology material where fixation times can vary from a few

Table 1. Typical levels of proteinase K digestion for *in situ* hybridization

Preparation	Fixation	Pre-treatment
Cell cultures cytospin	10% formal saline at 4°C for 20 min	no digestion
Frozen sections	10% formal saline or 4% paraformaldehyde at 4°C for 15 min	no digestion or mild proteinase K, 1 μg/ml for 30 min at 37°C
Cytoblock sections	10% formal saline at ambient temp. for 30 min	mild proteinase K, 1 μg/ml for 30 min at 37°C
Wax sections	4% paraformaldehyde at ambient temp. for 24 h	proteinase K, 2–20 μg/ml for 1 h at 37°C
Wax sections	10% formal saline at ambient temp. for 48 h	proteinase K, 2–50 μg/ml for 1 h at 37°C

days to many months (25). It should be noted that the optimum level of protease digestion must be determined empirically for each block, particularly with formalin-fixed material, and will always be below the level where there is morphological damage to the tissue. *Table 1* shows typical pre-treatment levels of proteolytic digestion for each type of *in situ* preparation.

Other pre-treatments are occasionally needed to prevent non-specific signals and improve the signal to noise ratio. Mild acid treatment with 0.1 M HCl reduces non-specific probe binding to tissues fixed in formal sublimate, and is thought to remove basic groups (26). However, subsequent proteinase K treatment must be modified as HCl pre-treatment aids digestion. Hot salt washes in 2 × SSC at 70°C also improves the signal and are thought to remove some of the secondary structure from RNA targets (26). When the probe binds to charged groups in the tissue, acetylation in acetic anhydride will block this non-specific background (27). It should be noted that with increasing levels of acetylation the hybridization signal will be reduced and, therefore, the optimum level of acetylation must be determined empirically for each tissue type. Finally post-fixation in 0.4% paraformaldehyde in 1 × PBS is required to re-cross-link any solubilized RNA to prevent its loss during subsequent incubations. *Protocol 7* describes a range of pre-treatments for formalin-fixed tissues and cells.

Protocol 7. Pre-treatments for formalin-fixed cells or tissues

Reagents

- xylene
- industrial methylated spirits (IMS)
- DEPC-treated H_2O (see *Protocol 1*)
- 2 × SSC (0.3 M sodium chloride, 30 mM sodium citrate) DEPC, pH 7.0
- proteinase K in 50 mM Tris–HCl, pH 7.6 (*Table 1*)
- 0.4% paraformaldehyde in 1 × PBS

Method

1. De-wax the sections to DEPC-H_2O as follows:
   - 2 changes of xylene for 5 min each
   - 2 changes of 99% IMS for 5 min each
   - 2 changes of 95% IMS for 5 min each
   - 2 changes of 75% IMS for 5 min each
   - DEPC-H_2O for 5 min, twice

2. Incubate the sections in a hot salt wash in 2 × SSC/DEPC-H_2O at 70°C for 10 min.

3. Incubate the sections in DEPC-H_2O at ambient temperature for 5 min.

4. Incubate in proteinase K in 50 mM Tris–HCl, pH 7.6. Concentration is dependent on tissue/cells and fixation. Usually 2 or 5 μg/ml is required at 37°C for 1 h.

5. Incubate the sections in two changes of DEPC-H_2O at 4°C for 5 min each.

6. Post-fix in 0.4% paraformaldehyde in 1 × PBS at 4°C for 20 min.

7. Incubate in two changes of DEPC-H_2O at 4°C for 5 min each.

Protocol 8. Acetylation of sections

Reagents

- industrial methylated spirits (IMS)
- 0.05% acetic anhydride (Sigma) in xylene
- DEPC-treated H_2O (*Protocol 1*)

Method

1. Proceed from *Protocol 7.*

2. Dehydrate the sections or cells using the following steps:
   - 2 changes of 95% IMS for 3 min each
   - 2 changes of 99% IMS for 3 min each

3. Prepare a fresh solution of 0.05% acetic anhydride in xylene.

4. Immerse the sections in acetic anhydride solution for 5 min.

5. Rehydrate the sections or cells using the following steps:
   - 2 changes of 99% IMS for 3 min each
   - 2 changes of 95% IMS for 3 min each
   - 2 changes of DEPC-H_2O for 5 min each.

3.7 Synthetic oligonucleotide probes for RNA *in situ*

For *in situ* hybridization, machine-made oligonucleotides are usually synthesized by phosphoramidite chemistry (28) over the range of about 25–30 bases in length and can be custom synthesized by a number of commercial suppliers.

A number of strategies for labelling synthetic oligonucleotides are available for non-radioactive probes. The haptens are covalently attached either enzymatically, using terminal deoxynucleotidyl transferase (29) (TdT) (*Protocol 9*) and a non-radioactive labelled nucleotide, or chemically, by the addition of a reactive 'amino-link' group at the 5' end (13, 30, 31) (*Protocol 10*). It is now possible to synthesize oligonucleotides with biotin, fluorescein, or other hapten groups directly coupled by a labelled phosphoramidite reaction (32). The sensitivity of oligonucleotides labelled with haptens depends upon the number of haptens coupled, the positioning of the haptens to avoid steric hindrance, and the number of separate oligonucleotide sequences used to make the probe. *Protocol 11* describes the use of spun columns to purify oligonucleotide probes.

The range of haptens available for non-radioactive *in situ* hybridization is shown in *Table 2*. Although the first non-radioactive label to be used was biotin, the alternative labels have the advantage that they are not endogenous in the tissue sections. Biotin is a co-factor in carboxylase enzymes and can also be present in cells associated with membrane transport proteins. High levels of endogenous biotin are present in liver, kidney, and proliferative sites. Currently, no reliable blocking techniques are available to prevent signal from endogenous biotin when a sensitive streptavidin detection system is used on these tissues.

Protocol 10 describes a method suitable for the 3' and 5' labelling of 5 μg quantities of oligonucleotides synthesized with a 5' aminolink group in place.

Table 2. The range of haptens available for non-radioactive *in situ* hybridization

Label	Detection method
Biotin	histochemical/immunochemical
Alkaline phosphatase	histochemical
5-bromo-deoxyuridine	immunochemical
Dinitrophenol	immunochemical
Digoxigenin	immunochemical
Fluorescein	fluorescent/immunochemical
Tetramethylrhodamine	fluorescent/immunochemical
AMCA (aminomethyl coumarin acetic acid)	fluorescent

Protocol 9. 3′ labelling of oligonucleotides using fluorescein or digoxigenin nucleotides

Equipment and reagents

- oligonucleotide cocktail
- 10 mM manganese chloride (Sigma)
- fluorescein-12-dUTP or digoxigenin-12-dUTP (Boehringer–Mannheim)
- terminal deoxynucleotidyl transferase (TdT)

and TdT buffer (Boehringer–Mannheim kit cat. no. 220582)
- 0.5 M EDTA, pH 8.0
- TE buffer, pH 8.0
- spun columns (*Protocol 11*)

Method

The following method is suitable for the 3′ labelling of 5 μg quantities of oligonucleotides.

1. Add in the following order:
   - sterile pure water (to a total volume of 100 μl)   w μl
   - 5 μg of oligonucleotide cocktail (crude or gel-purified full length)   x μl
   - 10 mM manganese chloride   10 μl
   - 10 nmol fluorescein-12-dUTP   y μl
     or 10nmol digoxigenin-12-dUTP
   - 50 U terminal deoxynucleotidyl transferase (TdT)   z μl
   - TdT buffer   20 μl

   Incubate for 2 h at 37°C.
2. After incubation, stop the enzyme activity by adding 5 μl of 0.5 M EDTA, pH 8.0.
3. Remove unincorporated nucleotides by using a 1 ml spun column equilibrated with 1 × TE, pH 8.0 (see *Protocol 11*).
4. Measure the volume of eluant and store at −20°C, at which temperature the probe should be stable for at least six months.

Protocol 10. 5′ and 3′ allyl amine labelling of oligonucleotides with fluorescein or digoxigenin

Equipment and reagents

- ultrapure sterile water
- oligonucleotide cocktail
- 10 mM manganese chloride (Sigma)
- 0.4 mM 5-(3-aminoallyl)-2′-deoxyuridine 5′-triphosphate (AA-dUTP) (Sigma)
- terminal deoxynucleotidyl transferase (TdT) and TdT buffer (see *Protocol 9*)
- 1 M EDTA, pH 8.0
- phenol/chloroform/isoamyl alcohol (25:24:1 v/v)
- spun column (see *Protocol 11*)

Protocol 10. *Continued*

Method

A. *3' Allyl addition reaction*

1. Add in the following order:
   - sterile ultrapure water (to a total volume of 100 µl)   x µl
   - 5 µg of oligonucleotide cocktail (crude or gel purified
     full length)[a]   y µl
   - 10 mM manganese chloride   10 µl
   - 10 mM AA-dUTP[a]   30 µl
   - 50 U TdT: 5 units/µg oligo   z µl
   - TdT buffer (*Protocol 9*)   20 µl

2. Incubate for 2 h at 37°C.

3. After incubation, stop the enzyme activity by adding 5 µl of 1 M EDTA, pH 8.0

4. Extract the TdT-labelled oligos using phenol as follows:
   - add 200 µl of phenol/chloroform:isoamyl alcohol; gently invert to mix then microcentrifuge for 1 min at 13 000 r.p.m.;
   - transfer the aqueous layer, taking care not to disturb the interface, into a new screw-capped Eppendorf tube;
   - remove traces of phenol and unincorporated nucleotides by using a 2 ml spun column equilibrated with 1 × TE, pH 8.0 (see *Protocol 11*).

B. *5' and 3' labelling* (32)

Reagent

- 1 M sodium borate buffer, pH 8.5
- 10 mg FITC isomer (Sigma) in 500 µl dimethyl formamide

- 1.5 mg digoxigenin-3-O-methylcarbonyl-ε-amino caproic acid-*N*-hydroxysuccinimide ester (DIG-NHS) ester (Boeringher–Mannheim 1333054) in 150 µl dimethyl formamide

Method

1. Add in order the following reagents:
   (a) *For FITC*
      - 1 M sodium borate buffer, pH 8.5.   25 µl
      - freshly prepared solution of 10 mg FITC isomer 1[b] in 500 µl dimethyl formamide   25 µl
      - sterile ultrapure water to   250 µl
   (b) *For digoxigenin*
      - 1 M sodium borate buffer, pH 8.5   25 µl

- freshly prepared solution of 1.5 mg DIG-NHS [b]
  in 150 μl dimethyl formamide   34 μl
- sterile ultrapure water   to 250 μl

2. Incubate overnight at room temperature.

3. Separate labelled oligonucleotides from the reaction solution through two, 2 ml volume spun columns equilibrated in 1 × TE buffer, pH 8.0 (see *Protocol 11*).

4. Store at 4°C.

---

[a] The molar ratio of oligonucleotide cocktail to nucleotides should be 1:25, i.e. 1 nmole of oligos to 25 nmoles of AA-dUTP.

[b] The volumes given represent a 50 molar excess of label to nucleic acid. Care should, therefore, be taken to check the concentration of each supplied batch of label and to adjust volumes if required

---

Protocol 11. Spun columns [a]

Equipment and reagents

- 1 ml disposable syringe
- sterile polyallomer wool or siliconized glass wool
- Sephadex G-50 (Pharmacia) pre-swollen in 1 × TE buffer, pH 8.0 (5 g in 100 ml)

A. Preparation

1. Plug the barrel of a 1 ml disposable syringe with sterile polyallomer wool or sterile, siliconized glass wool. Fill the syringe with a slurry of Sephadex G-50.

2. Let the TE buffer drain out and fill the syringe again with the Sephadex slurry. Continue adding Sephadex until the syringe is filled with gel when the TE buffer has drained out.

3. Insert the syringe into a polypropylene centrifuge tube. Centrifuge at 1600 *g* for 5 min in a bench centrifuge. Do not be alarmed at the appearance of the column. Continue to add Sephadex until the packed column volume after centrifugation is 0.9 ml.

4. Add 100 μl of TE buffer to the top of the column and re-centrifuge. Do this three times. Collect the last eluent in a de-capped Eppendorf tube and measure its volume, which should be 100 μl. If significantly different, add another 100 μl of TE buffer and spin again.

5. Cap the column and centrifuge tube with lab. film, store at 4°C or use immediately.

Protocol 11. *Continued*

B. *Use*

1. Place a 1.5 ml screw-cap Eppendorf tube below the spun column. Add 100 μl TE buffer to the top of the column and centrifuge at 1600 *g* for 5 min.

2. Measure eluant in the Eppendorf tube. If about 100 μl discard and proceed to next step. If significantly below 100 μl add a further 100 μl of TE buffer and re-centrifuge as in Step 1.

3. Place a new Eppendorf tube under the column. Add the probe solution to the top of the column and centrifuge at 1600 *g* for 5 min.

4. Measure the volume, cap the Eppendorf tube, label, and transfer to a −20°C freezer.

a Based on ref. 4.

3.8 Pre-hybridization, hybridization, and post-hybridization washes

Pre-hybridization of the specimen is necessary to prevent non-specific binding of the probe to the slide and tissue section. It also allows the hybridization conditions to be established prior to the addition of the probe. Pre-hybridization can be avoided by dehydrating the sections prior to the hybridization step and rehydrating the sections with hybridization buffer containing the probe.

Most studies on the effects of hybridization conditions on *in situ* hybridization signals show that similar factors are involved as those observed for filter hybridizations. Consequently the parameters influencing RNA *in situ* hybridization may be empirically determined in the same manner as for Northern blotting experiments. The stability of any given oligonucleotide:RNA hybrid is a function of its length, base composition, and hybridization conditions. Although a 20-base oligomer can be specific for the human genome, longer oligonucleotides show increased hybridization efficiency. Therefore, by synthesizing 25–30 base oligomers a compromise can be achieved between the ideal length for hybrid stability, specificity, cost of synthesis, cost of purification, and tissue penetration efficiency. The sequences chosen for each oligonucleotide should be on average above 50% G + C content for increased hybridization efficiency but below 70% G + C content to avoid non-specific cross-hybridization with other G + C (33) rich RNA.

The thermal stability of the *in situ* hybrids is characterized by the mean thermal denaturation temperature (T_m), the temperature at which 50% of the hybrids dissociate. Comparative studies of *in situ* hybridization and solution or filter hybridization have shown that the same parameters control hybridization efficiency and that the *in situ* T_m of heterologous duplexes is 5°C below the solution or filter T_m (26). RNA:DNA hybrids are more stable than DNA:DNA hybrids and so dissociate at higher temperatures. Typically the difference in T_m is 11 °C (34). Therefore, theoretically the T_m of any *in situ* hybridization reaction can be calculated using the following relationship which has been derived by combining several results for DNA:DNA and DNA:RNA hybrids (26, 34, 35):

$$T_m = 87.5 + 16.6(\log M) + 0.41 (\%G + C) - 0.61 (\%formamide) - 650/n,$$
(1)

where M is the molarity of the monovalent cation and (%G +C) is the percentage of guanine and cytosine residues in the probe DNA. The reduction in T_m by formamide is not linear for RNA:DNA hybrids and is not valid above 50% formamide (36). The last term corrects for probe length, where n is the length of the probe in bases. The concentration of oligonucleotides in the hybridization buffer depends upon the number of separate oligomers in the probe cocktail. Concentrations approaching 50 ng/oligomer/ml give good signal to noise ratios on most tissues. Total probe concentrations above 1–2 µg/ml should be avoided as this will increase the background above acceptable levels.

Blocking of non-specific binding of the probe is achieved using macromolecules in the hybridization buffer at a concentration of 100–200 mg/ml. Dextran sulfate at 10% is an essential component, not only because it improves signal due to its accelerating effects but also it has considerable blocking activity. In the absence of dextran sulfate probes have a tendency to bind non-specifically to eosinophilic cells, but its addition blocks this effect. Other polymers, such as PEG 6000, cannot be substituted because it does not suppress this artefact. Other macromolecules used for blocking include: heterologous DNA/RNA (salmon or herring sperm DNA, transfer RNA) and Denhardt's solution. *Protocol 12* describes the hybridization conditions needed for RNA *in situ* using oligonucleotide probes.

Once hybridization has proceeded to completion, it is necessary to wash off any mismatched or non-specifically bound probe. By controlling the temperature and ionic strength of the washing solutions, it is possible to promote dissociation of mismatched probe/target sequences and to obtain higher specificity of the final detected signal.

Protocol 12. Pre-hybridization, hybridization, and post-hybridization

Reagents[a]

- pre-hybridization solution (see Step 1)
- hybridization solution (see Step 2)
- probe
- 2 × SSC: 0.3 M sodium chloride, 30 mM sodium citrate
- 2 × SSC/30% formamide

- 10 × PE: 0.5 M Tris–HCl, pH 7.5, 1% sodium pyrophosphate; 2% polyvinylpyrrolidone (mol. wt 40000); 2% Ficoll (mol. wt 400000); 50 mM EDTA. Dissolve by heating to 65 °C. Once dissolved hold at this temperature for 15 min. Store at 4 °C.

Protocol 12. *Continued*

Method

1. Prepare pre-hybridization solution from stock solutions:

		Final concentration
• 5 M NaCl	120 μl	0.6 M NaCl
• 10 × PE [a]	100 μl	1 × PE
• 50% dextran sulfate	200 μl	10% dextran sulfate
• 10 μg/ml ssDNA	15 μl	150 μg/ml ssDNA
• formamide	300 μl	30% formamide
• DEPC-treated H$_2$O	265 μl	
• Total volume	1000 μl	

2. Prepare hybridization solution as follows:
   • Add all pre-hybridization reagents, except the DEPC-treated H$_2$O to an Eppendorf tube. Then add the probe cocktails at double their final concentration and finally make up to the required volume by adding DEPC-treated water.

3. Place the slides into a humidity chamber [b] and cover each section with pre-heated pre-hybridization solution (25–50 μl per section). Incubate the humidity chamber in an air incubator at 37°C for 1 h.

4. Leave the pre-hybridization solution on the slide then add an equal volume of hybridization solution. Add a siliconized coverslip to ensure even distribution of the probe and retention during incubation. Hybridize in a humidity chamber with a grease-sealed lid at 37°C overnight.

5. Remove coverslips by rinsing slides in 2 × SSC at ambient temperature.

6. Post-hybridize sections in Coplin jars in two changes of 2 × SSC/30% formamide pre-warmed to 37°C in a waterbath for 10 min each.

[a] All solutions should be made up in DEPC-treated ultrapure water.
[b] An alternative incubation system such as the OmniSlide, Hybaid, can be used. This system has a humidity chamber over a temperature-controlled heating block.

3.9 Detection systems

Non-radioactive labels are generally detected by anti-hapten monoclonal or polyclonal antibodies coupled to various visualization systems such as alkaline phosphatase, peroxidase, immunogold with silver enhancement or fluorescence (37–39). For digoxigenin and fluorescein labelled probes, systems employing FAB fragment polyclonal primary antibodies conjugated to alkaline phosphatase are very sensitive. Alternatively, oligonucleotides can be labelled with fluorochromes or alkaline phosphatase and visualized by direct methods.

In general, alkaline phosphatase is the enzyme of choice. This can be detected by incubating with 5-bromo-4-chloro-3-indolyl phosphate (BCIP enzyme substrate) and Nitro blue tetrazolium (NBT chromogen) or using the Naphthol phosphate (BI) substrate and Fast Red TR chromogen method (*Protocol 13*). Both systems give good resolution and signal strength, but slides have to be mounted in aqueous mountants such as gylcerol to prevent solubilization of the end product. If the Naphthol phosphate (MX) substrate is used with the Fast Red TR chromogen a fluorescent product is precipitated at the site of the enzyme which can be combined with fluorescein.

Protocol 13. Antibody detection systems for oligonucleotides labelled with fluorescein or digoxigenin

Reagents

A. *BCIP–NBT visualization*

- blocking solution: 3% BSA, 0.1% Triton-X-100 in TBS. Filter before use
- anti-digoxigenin alkaline phosphatase (Boehringer–Mannheim)
- anti-fluorescein alkaline phosphatase (Boehringer–Mannheim)
- TBS
- NBT: 75 mg/ml in 70% dimethyl formamide
- BCIP: 50 mg/ml in dimethyl formamide
- 1 M levamisole
- BCIP–NBT substrate buffer: 10 ml 1 M Tris, pH 9.0; 5 ml 1 M $MgCl_2$; 2 ml 1 M NaCl; 83 ml ultrapure water
- Mayers Haematoxylin (optional)
- mountant: Apathys or Glycergel

The following detection systems should be carried out at ambient temperature.

1. Incubate the slides in blocking solution for 5 min.

2. Cover the sections with 100–200 μl of either anti-digoxigenin alkaline phosphatase conjugate (1:600) or anti-fluorescein alkaline phosphatase conjugate (1:600) diluted in blocking solution for 30 min.

3. Wash the slides in two changes of TBS for 5 min each.

4. Wash slides in BCIP–NBT substrate buffer for 5 min.

5. Make up BCIP–NBT substrate solution as follows:
   - NBT   8 μl
   - BCIP   8 μl
   - 1 M levamisole   1 μl
   - BCIP–NBT substrate buffer 1 ml

6. Apply 200 μl per section. Coverslip the sections and incubate in the dark—check microscopically until maximum signal is obtained before background develops, from 1 h to overnight.

7. Wash in running tap water for 5 min.

8. Counterstain, if required, for > 1 sec in Mayers Haematoxylin.

9. Wash in running tap water for 5 min.

10. Mount in Apathys or Glycergel.

Protocol 13. *Continued*

B. *Fast Red visualization*

Equipment and reagents

- veronal acetate substrate buffer, pH 9.2: 30 mM sodium acetate trihydrate; 30 mM sodium barbitone; 100 mM sodium chloride; 50 mM magnesium chloride hexahydrate
- Fast Red TR Salt (Sigma)

- naphthol ASBI or ASMX Phosphate (Sigma)
- dimethyl formamide
- Mayers haematoxylin (optional)
- Whatman filter paper
- mountant: Apathys or Glycergel

1. Follow steps A1–5 above.
2. Wash slides in veronal acetate substrate buffer, pH 9.2, for 5 min.
3. Make up Fast Red substrate solution in fume cupboard as follows:
   - Fast Red TR salt   150 mg
   - Naphthol ASBI phosphate or Naphthol ASMX phosphate   150 mg (pre-dissolve in 1 m dimethyl formamide)
   - 1 M levamisole   72 mg
   - veronal acetate substrate buffer   300 ml

   Filter through Whatman filter paper before incubating with slides.
4. Incubate the sections submerged in Fast Red substrate solution for 1 h.
5. Wash in running tap water for 5 min.
6. Counterstain, if required, for < 1 sec in Mayers haematoxylin.
7. Wash in running tap water for 5 min.
8. Mount in Apathys or Glycergel.

The simultaneous detection of several probes in cells or tissue sections can be achieved with probes bearing different labels. These are demonstrated with different distinguishable visualization systems. The best resolution is achieved by using two fluorochromes, for example fluorescein (FITC) in combination with rhodamine (TRITC) (40). Recently, three DNA probes have been simultaneously detected using FITC, TRITC, and AMCA as the fluorescent labels (41). Peroxidase and alkaline phosphatase have also been used to detect two different target sequences using biotin and digoxigenin labels (42, 43). For combined immunocytochemistry and *in situ* hybridization similar strategies are required to give the best resolution (see Chapter 5 of this volume).

4. Applications

4.1 Detection of heterogeneous nuclear RNA (hnRNA)

HnRNAs are the primary transcripts produced in the nucleus prior to RNA processing. Information about the site of hnRNA expression may be used to determine which cells are transcribing a gene. However, this information may not correlate with cells that express the gene product. For example interleukin-1β hnRNA expression can be detected in macrophages, but this RNA is not necessarily processed to mature mRNA and, therefore, does not produce interleukin-1β protein. The heterogeneous RNA is broken down rapidly in the nucleus. Probes to detect nuclear expression of genes can be designed to intron sequences which are only present in the nucleus. Introns are removed during RNA processing by splicing enzymes and, therefore, *in situ* signals present only in the nucleus of the cell indicate that transcription may have just been initiated, or that cells are receiving the appropriate signals for transcription to start but inappropriate signals for RNA processing.

4.2 Messenger RNA

Mature messenger RNA can be detected in the cytoplasm of cells using probes to exon sequences or exon boundaries. The success of mature mRNA detection depends upon the copy number of the steady-state mRNA in the cell cytoplasm as mature mRNA is degraded, as genes are expressed and mRNAs have different half-lives. The use of probes to detect mRNA rather than detecting the gene product to establish the site of synthesis can be useful in a number of circumstances, for example:

(a) When the gene product is rapidly degraded in the cell or rapidly released from the cell.

(b) When the gene product is present in the tissues as extracellular protein. In this situation immunohistochemical methods will fail to localize the cells expressing the gene product (see *Plate 5*) (44).

(c) When the gene product is mutated and, therefore, the epitope in the mutated protein may not be recognized by antibodies.

(d) Antibodies are not available for the gene product because of self-antigens.

(e) Antibody methods fail because antigens are not preserved in archival formalin-fixed paraffin-embedded tissue.

In situ hybridization has some advantages over immunocytochemistry; it can establish the site, the temporal distribution, or the morphological distribution of expression of RNA within tissues. Frequently the half-life of mRNA differs from its protein product, and RNA detection may, therefore, give information about the steady-state production of a new protein, whereas detection of protein reflects both newly synthesized and stored protein. One example where the rate of synthesis is important is the measurement of proliferation by detecting histone mRNA. The level of histone mRNA increases by 30–50 fold during 'S phase'. Therefore, *in situ* detection of histone mRNA shows cells within S phase of the cell cycle (see *Plate 6*) (49). The detection of other mRNAs within tissues, particularly in pathological circumstances, can indicate how a cell is responding to environmental changes. Therefore, *in situ* hybridization can be usefully applied to study adaptive responses, as in stress response where genes such as ubiquitin, haemoxigenase, 70 kDa heat-shock protein are induced during cell stress. *Plate 7* shows the expression of glyceraldehyde-3-phosphate dehydrogenase mRNA in skeletal muscle indicating the level of enzyme in different fibre types.

4.3 Detection of RNA in organelles

The detection of RNA in organelles (45) can be useful in order to:

(a) demonstrate the presence of organelles

(b) establish the condition of organelles

(c) investigate the relationship between organelle proteins expressed from either the nuclear or organelle genome.

Plate 8 shows the detection of the rRNAs within mitochondria in a paraffin section of human kidney. This probe can be used to demonstrate RNA preservation because the mitochondrial RNAs are quickly destroyed in tissues that have been inadequately collected and fixed. The probe also shows the distribution of mitochondria in different cell types.

4.4 Viral RNAs

One of the most important areas for *in situ* hybridization has been the detection of viruses (46, 47). The advantage of detecting gene expression from the viral genome in cells rather than the genome itself is that it can make use of higher copy numbers of targets to improve the sensitivity and give information about whether the infection is latent or lytic (48). It also allows studies of tissue trophism, as well as information about the pathogenicity of the virus, and probes can be designed to different viral strains. Probes to mRNA from viral sequences or organelle sequences can be derived from exons which can be identified from the genomic sequence. It is important to remember that transcription of double stranded DNA viruses is often bi-directional and, therefore, the use of antisense controls may be inappropriate. *Plates 9* and *10* show the detection of human viruses. Other applications in pathology are reviewed in Warford and Lauder (50).

Acknowledgements

I am grateful to Mr A. Warford for help with the protocols and Dr D. Hilton for the information and figures on coxsackie virus.

References

1. Darnell, J. E. (1982). *Nature*, **297**, 365.
2. Moore, P. B. (1988). *Nature*, **331**, 223.
3. Maniatis, T. and Reed, R. (1987). *Nature*, **325**, 673.
4. Sambrook, J., Fritsch, E. F., and Maniatis, T. (ed.) (1989). *Molecular cloning: a laboratory manual* (2nd edn.), Vols 1, 2, and 3. Cold Spring Harbor Laboratory, New York.
5. Langer, P. R., Waldrop, A. A., and Ward, D. C. (1981). *Proc. Natl Acad. Sci. USA*, **78**, 6633.
6. Niedobitdk, G., Finn, T., Herbst, H., Borbhoft, G., Gerdes, J., and Stein, H. (1988). *Am. J. Pathol.*, **131**, 1.
7. Herrington, C. S., Burns, J., Graham, A. K., Evans, M. F., and McGee, J. O'D. (1989). *J. Clin. Pathol.*, **42**, 592.
8. Landegent, J. E., Jansen, in de Wal N., Baan, R. A., Hoeymakers, J. H. J., and Van der Ploeg, M. (1984). *Exp. Cell Res.*, **153**, 61.
9. Tchen, P., Fuchs, R. P. P., Sage, E., and Leng, M. (1984). *Proc. Nati Acad. Sci. USA*, **81**, 3466.
10. Hopman, A. H. N., Wiegant, J., and Duijn, P., van. (1986). *Histochemistry*, **84**, 169.
11. Morimoto, H., Monden, T., Shimano, T., Higashiyama, M., Tomita, N., Murotani, M., Matsuura, N., Okuda, H., and Mori, T. (1987). *Lab. Invest.*, **57**, 737.
12. Forster, A. C., McInnes, J. L., Skingle, D. C., and Symons, R. H. (1985). *Nucl. Acids Res.*, **13**, 745.
13. Cook, A. F., Voucolo, E., and Brakel, Ch., L. (1988). *Nucl. Acids Res.*, **16**, 4077.
14. Coghlan, J. P., Aldred, P., Haralambidis, J., Niall, H. D., Penschow, J. D., and Tregear, G. W. (1985). *Anal. Biochem.*, **149**, 1.
15. Lewis, M. E., Sherman, T. G., and Watson, S. J. (1985). *Peptides*, **6**, 75.
16. Pringle, J. H., Primrose, L., Kind, C. N., Talbot, I. C., and Lauder, I. (1989). *J Pathol.*, **158**, 279.
17. Lawrence, J. B. and Singer, R. H. (1985). *Nucl. Acids Res.*, **13**, 1777.
18. Urieli-Skoval, S., Meek, R. L., Hanson, R. H., Ferguson, M., Gordon, D., and Benditt, E. P. (1992). *J. Histochem. Cytochem.*, **40**, 1879.
19. Singer, R. H., Lawrence, J. B., and Villnave, C. (1986). *Biotechniques*, **4**, 230.
20. Moench, T. R. (1987). *Mol. Cell. Probes*, **1**, 195.
21. Hoefler, H., Chiders, H., Montminy, M. R., *et al.* (1986). *Histochem. J.*, **18**, 597.
22. Huang, W. M., Gibson, S. J., Facer, P. *et al.* (1983). *Histochemistry*, **77**, 275.
23. Van Prooijen-Kneght, A. C., Raap, A. K., Van der Burg, M. J. M, *et al.* (1983). *Histochem. J.*, **14**, 333.
24. Bancroft, J. D. and Stevens, A. (ed.) (1982). *Theory and practice of histological techniques*. Churchill Livingstone, Edinburgh.
25. Pringle, J. H., Ruprai, A. K., *et al.* (1990). *J. Pathol.*, **162**, 197.
26. Cox, K. H., DeLeon, D. V., *et al.* (1984). *Devel. Biol.*, **101**, 485.
27. Hayashi, S., Gillam, I. C., Delaney, A. D., and Tener, G. M. (1978). *J. Histochem. Cytochem.*, **26**, 677.
28. Beaucage, S. L. and Caruthers, M. H. (1981). *Tetrahedron Lett.*, **22**, 1859.
29. Deng, G. and Wu, R. (1981). *Nucl. Acids Res.*, **9**, 4173.

30. Applied Biosystems Ltd. (1986). Model 380A/B DNA Synthesizer. *User bulletin*, No. 30.
31. Costello, S. M., Felix, R. T., and Giese, R. W. (1979). *Clin. Chem.*, **25**, 1572.
32. Chu, B. C. F. and Orgel, L. E. (1985). *DNA*, **4**, 327.
33. Agarwal, K. L., Brunstedt, J., and Noyes, B. E. (1981). *J. Biol. Chem.*, **256**, 1023.
34. Anderson, M. L. M. and Young, B. D. (1985). In *Nucleic acid hybridization: a practical approach* (ed. B. D. Hames and S. J. Higgins), pp. 73–111. IRL Press, Oxford.
35. Howley, P. M., Israel, M. F., Law, M.-F., and Martin, M. A. (1979). *J. Biol. Chem.*, **254**, 4876.
36. Casey, J. and Davidson, N. (1977). *Nucl. Acids Res.*, **4**, 1539.
37. De Jong, A. S. H., Van Kessel-Van Vark, M., and Raap, A. K. (1985). *Histochem. J.*, **17**, 1119.
38. Pringle, J. H., Ruprai, A. K., *et al.* (1990). *J. Pathol.*, **162**, 197.
39. Harper, S. J., Pringle, J. H., Gillies, A., Allen, A. C., Layward, L., Feehally, J., and Lauder, I. (1991). *J. Clin. Pathol.*, **45**, 114.
40. Cremer, T., Tessin, D., Hopman, A. H. N., and Manuelidis, L. (1988). *Exp. Cell Res.*, **176**, 199.
41. Nederlof, P. M., Robinson, D., Abuknesha. R., Wiegant. J., Hopman, A. H. N., Tanke, H. J., and Raap, A. K. (1989). *Cytometry*, **10**, 20.
42. Emmerich, P., Loos, P., Jauch, A., Hopman, A. H. N., Wiegant, J., Higgins, M., White, B. N., van der Ploeg, M., Cremer, C., and Cremer, T. (1989). *Exp. Cell Res.*, **181**, 126.
43. Herrington, C. S., Burns, J., Bhatt, B., Graham, A. K., and McGee, J. O'D. (1989). *J. Clin. Pathol.*, **42**, 601.
44. Kendall, C. H., Roberts, P. A., *et al.* (1991). *J. Pathol.*, **165**, 111.
45. McCabe, J. T., Morrell, J. I., and Ivell, R. (1986). *J. Histochem. Cytochem.*, **34**, 45.
46. Grody, W. W., Cheng, L., and Lewin, K. J. (1987). *Hum. Pathol.*, **18**, 535.
47. Lowe, J. B. (1986). *Clin. Chim. Acta*, **157**, 1.
48. Pringle, J. H., Barker, S., *et al.* (1992). *J. Pathol. (Suppl.)*, **167**, 133A.
49. Pringle, J. H., Baker, J., *et al.* (1993). *J. Pathol. (Suppl.)*, **169**, 144A.
50. Warford, A. and Lauder, I. (1991). *J. Clin. Pathol.*, **44**, 177.

5

Combination of non-radioactive *in situ* hybridization and immunocytochemistry

H. MULLINK, W. VOS, N. M. JIWA, A. HORSTMAN,
E. RIEGER, and C. J. L. M. MEIJER

1. Introduction

In situ hybridization is a powerful tool for the detection and localization of nucleic acid sequences in cells and tissue sections. In this way, the presence of specific DNA fragments as well as (messenger) RNA products can be demonstrated with preservation of morphology (1–3). As morphology may be impaired during *in situ* hybridization, additional information regarding the cells involved may be required, depending on the scientific or diagnostic problem being studied. Since a particular cellular phenotype is characterized by the presence of specific proteins or antigens, immunocytochemical staining is required to reveal it. This phenotype can be studied by careful comparison of adjacent tissue sections stained by *in situ* hybridization and immunocytochemistry respectively. However, a more accurate and easier method is to analyse both simultaneously in one single, double-stained section. Moreover, only this approach allows the study of individual cells.

Most techniques described for the combination of *in situ* hybridization and immunocytochemistry make use of radioactive probes, and often require sophisticated procedures for fixation and tissue processing (4–6). The use of non-radioactive procedures has considerable advantages; they are considerably shorter procedures and do not need special laboratory facilities. For these reasons radioactive probes used in *in situ* hybridization are being supplanted more and more by their non-radioactive counterparts. The procedures for non-radioactive *in situ* hybridization have improved in reliability and sensitivity, and can often be combined with immunohistochemistry, provided certain conditions are kept in mind (7–9).

In this chapter we will restrict ourselves to combinations of DNA or RNA *in situ* hybridization with immunohistochemistry, in which both tissue elements of interest are revealed finally by means of immunoenzyme detection methods.

If *in situ* hybridization has to be combined with immunohistochemistry there are other factors, apart from preservation of morphology and the sensitivity of nucleic acid detection, which have to be taken into account. These are, for example, the preservation of tissue epitopes, the activity of enzyme labels, as well as the stability of the coloured cytochemical precipitates and of the antigen–antibody complexes during the subsequent stages of a combined procedure. Therefore, the assessment of the sequence in which both techniques should be performed is of great importance. In this respect the conditions for DNA and RNA *in situ* hybridization differ considerably.

The first condition that has to be fulfilled in double-stained preparations is that both elements of interest must be distinguished clearly from each other. This means that the chromogens used in both immunoenzyme detection methods should be of contrasting colours. Because double hybrido- and immunocytochemistry can only be performed by sequential, methods, there are two possibilities: (a) the same enzyme label is used for both parts of the staining combination but with different chromogens, or (b) two different enzyme labels are used with their respective, selected chromogens. Although theoretically many possible combinations exist (10), this number is considerably restricted in reality. First, many chromogens give staining which is too weak to be used in double-staining experiments, while others, which in fact represent the majority, do not withstand the physical or chemical conditions to be used in the second part of the procedure. There may also be a difference between various chromogen products in their ability to mask tissue targets of interest and render them inaccessible to the reagents used in the subsequent stages of the procedure. Still other chromogens are problematic because they have affinity for nucleic acid probes or have a special tendency to lead to background staining.

2. Combination of DNA *in situ* hybridization (DISH) with immunocytochemistry (IC)

The main application of DISH lies principally in two fields: (a) the detection of viral DNA in cytological preparations and tissue sections, and (b) the localization of nuclear sequences (genes) on chromosomal or interphase preparations. This chapter will be restricted to the detection of viral DNA sequences.

DISH allows the study of the distribution of infected cells within the tissue, often before the corresponding viral antigens can be detected. However, in many instances more specific information regarding the infected cell types would be of great advantage. Because the DISH procedure requires a compromise between morphological appearance and sensitivity of the DNA detection, the characterization of infected cells is often difficult. Moreover, the phenotype of morphologically identical cells is often different. Distinction may, however, be essential, for example in the study of viral carcinogenesis or lymphomagenesis (11, 12).

However, when DISH has to be combined with IC the essential aim is the combination of sensitive DNA detection on the one hand with preservation of the morphology and tissue antigens, antigen/antibody complexes, activity of enzyme labels, and stability of coloured pigments on the other hand. In other words, what is the best sequence for performing the different steps in the combined procedure in order to get the best compromise between these more or less contradictory experimental conditions?

Although double DISH–IC stainings have been described in which the *in situ* hybridization could be performed first because very stable antigens were involved (13), in the majority of cases the other sequence, (immunostaining followed by *in situ* hybridization), is preferred. Pre-treatment for DISH often involves proteolysis, and DNA requires strong denaturating conditions such as high concentrations of formamide as well as incubation at high temperature. The demonstration of vulnerable tissue elements is often no longer possible after these steps. In particular, those antigenic epitopes which can normally only be demonstrated on frozen sections are generally destroyed by these conditions. It is principally for this reason that the sequence for combined staining by DISH and IC should start with immunostaining which should run to completion, including colour development. Antigen–antibody bonds created in the course of these immuno-incubations and the enzyme activities of the conjugates used are generally not resistant to the harsh conditions necessary for DISH. This brings us to the next problem encountered in combined staining, namely the stability of the coloured pigments created during the immunostaining. Only a few of the known chromogens can withstand the reagents, solvents, and physical conditions which have to be used in the subsequent DISH procedure. Such chromogens are, for example diaminobenzidine (DAB) for peroxidase, Naphthol ASBI phosphate with diazotized New Fuchsin for alkaline phosphatase, and BCIG with ferri- and ferrocyanide for beta-galactosidase.

2.1 Preparation of microscope slides and coverslips

Proper preparation of microscope slides is important to prevent the tissue sections from being detached from the slides. The severe conditions necessary for DISH demand special coating of the slides to keep the tissue sections in place. In our experience, the use of 3-aminopropyltriethoxysilane (APES) coated slides (14) (*Protocol 1*) is effective in almost all experiments involving DISH, including combinations of DISH and IC on paraffin sections as well as on frozen sections. For adhesion of frozen tissue sections, poly-L-lysine coated glass slides can also be used (*Protocol 2*). Coverslips may be either acid treated or coated in APES (*Protocol 3*).

113

Protocol 1. Preparation of microscope slides for paraffin sections

Equipment and reagents

- 5–7% Lipsol
- 3-aminopropyltriethoxysilane (APES) (Sigma)
- acetone
- 3% glutaraldehyde
- slides in metal racks

Method

1. Place the slides in metal racks (30 slide capacity).
2. Clean the slides by immersing them in 5–7% Lipsol, or other cleaning liquid, for 18 h at room temperature.
3. Rinse twice in distilled water for 5 min each time.
4. Immerse the slides in acetone for 5 min and air-dry.
5. Immerse the slides in 3% (v/v) APES in acetone[a] for 20 min.
6. Rinse in acetone for 5 min.
7. Rinse in distilled water for 5 min.
8. Dry the slides in an oven at approx. 40°C overnight.
9. Store the slides in dust-free boxes. Before use the slides have to be activated.[b]
10. Incubate them in 3% glutaraldehyde for 5 min.
11. Rinse in tap water for 3 min.
12. Rinse twice in distilled water for 3 min each time.
13. Air-dry the slides.

[a] APES solution must be handled wearing gloves and in a fume cupboard. The APES solution must be made fresh every day.
[b] If the slides are kept for two weeks or longer, activation should be repeated.

Protocol 2. Preparation of poly-L-lysine (PLL) coated slides for frozen sections

Equipment and reagents

- distilled water
- poly-L-lysine (PLL) (mol.wt > 200 000, Sigma)
- Slides

Method

1. Dissolve PLL at a concentration of 0.1% (w/v) in distilled water.[a]
2. Coat the slides thinly by smearing the PLL solution over the slide. Mark the coated side of the slide with a pencil or glasswriter.

3. Air-dry the slides with a hair-drier or in a 40°C dry oven.
4. Store the slides in dust-free boxes until use.

> [a] Alternatively, PLL is dissolved at 1% (w/v) in distilled water and frozen in aliquots of, e.g. 0.5 ml. Before use one aliquot is thawed and diluted to 0.1% with distilled water.

Protocol 3. Preparation of coverslips [a]

Equipment and reagents

- 1 M HCl
- 95% ethanol
- coverslips

Method

1. Wash the coverslips in 1 M HCl.
2. Rinse them in tap water, then distilled water for 5 min each.
3. Wash twice in 95% ethanol for 15 min each time.
4. Air-dry before use. [b]
5. Store in a dust-free place.

> [a] Coverslips may also be prepared according to *Protocol 1*.
> [b] For RNA *in situ* hybridization: heat overnight in a dry oven at 180°C.

2.2 Preparation of specimens

2.2.1 Fixation

In experimental work, for paraffin as well as frozen sections, the choice of fixative is dictated in the first instance by the nature of the antigen to be detected, hence many possibilities exist. For each particular situation it should be determined whether the best fixative for antigen preservation is also compatible with DISH. However, there are some fixatives which have been shown to be suitable in many instances where IC and DISH have to be combined. Sometimes 4% paraformaldehyde in PBS is useful (*Protocol 4*), while in other situations paraformaldehyde–lysine–periodic acid (PLP) is preferred (*Protocol 5*). Acetic acid–paraformaldehyde is often suitable (*Protocol 6*) for cytospin- or cytological preparations.

In many cases, however, such as in diagnostic pathology, routinely one is dependent mostly on formalin-fixed and paraffin-embedded tissue. For retrospective analysis of archival material, apart from formalin-fixed specimens, Bouin's or formol-sublimate fixed tissue may be also encountered, often having variable fixation times. These tissues are often variable in their suitability for DISH and/or IC, and in such cases the protocols have to be modified, with respect to tissue pre-treatment. Here we describe the preparation of some fixatives that have been successful in different situations in which IC as well as DISH were used (*Protocols 4–6*).

Protocol 4. Preparation of 4% paraformaldehyde in PBS

Equipment and reagents

- paraformaldehyde
- phosphate buffered saline (PBS), pH 7.4
- 5 M NaOH
- 5 M HCl
- Whatman filter paper or Millipore filter

Method

1. Mix 4 g paraformaldehyde in 100 ml PBS.

2. Heat the solution to 60–70°C, add a few drops of 5 M NaOH, and stir until dissolved.

3. Control and adjust the pH to 7.4 with 5 M HCl.

4. Filter the solution through Whatman filter paper or use a Millipore filter.

5. Store the solution for up to one week at 4°C.

Protocol 5. Preparation of paraformaldehyde–lysine–periodate fixative (PLP)[a]

Reagents

- 0.2 M L-lysine
- 0.2 M phosphate buffer, pH 7.4
- paraformaldehyde
- 5 M NaOH
- sodium periodate

Method

1. Prepare solution A by mixing the following:
   - 0.2 M L-lysine   112 ml
   - 0.2 M phosphate buffer, pH 7.4   75 ml
   - H_2O 13 ml
   - adjust the pH to 7.4.

2. Prepare solution B by mixing:
   - paraformaldehyde   6 g
   - H_2O 80 ml
   - dissolve by adding 4–6 drops of 5 M NaOH to the solution, and stir

---

- add phosphate buffer, pH 7.4   5 ml
- adjust the pH to 7.4
- add water   to 100 ml
- and sodium periodate   640 mg
- stir until dissolved.

3. Mix solutions A and B for 300 ml PLP fixative.
4. Keep the solution at 4°C for up to one week.

[a] Based on ref. 15.

Protocol 6. Preparation of acetic acid–paraformaldehyde

Reagents

- 4% paraformaldehyde in PBS (see *Protocol*
  4)
- glacial acetic acid

Method

1. Take 10 ml of 4% paraformaldehyde solution in PBS.
2. Add 0.5 ml glacial acetic acid.
3. Keep solution at 4°C for up to one week.

2.2.2 Preparation of paraffin sections

Protocol 7 should be used for the preparation, de-waxing, and de-pigmentation of paraffin sections.

Protocol 7. Preparation and de-waxing of paraffin sections

Equipment and reagents

- APES-coated slides (see *Protocol 1*)
- xylene
- ethanol series
- Lugol's iodine
- sodium thiosulfate
- methanol
- H_2O_2

Method

1. Cut 4–5 μm thick sections and place on to APES-coated slides.
2. Blot the excess water and dry the sections on a hot plate at 40–50°C for 10–15 min.
3. Keep the slides at 37°C overnight. [a]

Protocol 7. *Continued*

4. De-wax the sections in xylene, using at least two changes for 15 min each.

5. Hydrate the sections by immersion in 98, 98, 90, 70, and 50% ethanol and then water, each for 1 min.

6. Incubate the sections in 2% Lugol's iodine in distilled water for 5 min.[b]

7. Wash them in running tap water for 3 min.

8. Incubate the sections in 2% (w/v) sodium thiosulfate, dissolved in distilled water, for 2 min.

9. Wash them in running tap water for 3 min.

10. Incubate the sections in 100% methanol, containing 0.3% H_2O_2 for 20 min.[c]

11. Rinse in tap water for 5 min.

[a] After Step 4 the sections can be kept for several months, at least for DISH, but only a limited time (some weeks) for RNA *in situ* hybridization.
[b] Steps 6–8 are necessary when silver or gold/silver based detection methods are to be used.
[c] Steps 10 and 11 must be performed when any immunoperoxidase detection methods are to be used.

2.2.3 Preparation of frozen sections

Protocol 8 should be used for the preparation of frozen sections.

Protocol 8. Preparation of frozen tissue sections

Equipment and reagents

- snap-frozen tissue blocks
- PLL- or APES-coated glass slides (*Protocols 1 and 2*)
- PBS
- fixative of choice: acetone, PLP (*Protocol 5*), or 4% paraformaldehyde in PBS (*Protocol 4*)

Method

1. Cut 4–6 μm thick sections from snap-frozen tissue blocks on a cryostat microtome.

2. Stick the sections on PLL- or APES-coated glass slides (*Protocols 1 and 2*).

3. Air-dry the sections for about 30 min.

4. Rinse the sections with PBS.

5. Fix the sections in the fixative of choice:
   - pure, dry acetone for 10 min at 4°C, then air-dry for 1 min;

or

- PLP (*Protocol 5*) for 10 min at 4°C, then wash in PBS three times for 5 min each;

or

- 4% paraformaldehyde in PBS (*Protocol 4*) for 10 min, then wash three times in PBS for 5 min each.

2.3 Immunocytochemistry

For the combination of immunohisto- or cytochemistry with DISH immunostaining should preferably be performed first. As mentioned above, DISH can only be performed first when the antigen to be demonstrated is extraordinarily stable and can be immunostained with very low background. Therefore, this chapter will be confined to the first possibility.

When IC is performed first, the choice of method depends on the properties of the pigment that will be formed by immunostaining. Its solubility and stability, its permeability for the probe, and its adhesive properties to the probe are important. Although immunoperoxidase (IPO), as well as immuno-alkaline phosphatase (IAP) and immuno-beta-galactosidase (IBG) methods are available, only a restricted number of available chromogens is suitable.

2.3.1 Immunoperoxidase staining of paraffin sections

For paraffin sections 'ordinary' IPO methods are adequate in most cases, using diaminobenzidine (DAB) and hydrogen peroxide as substrate. Common indirect, peroxidase–anti-peroxidase (PAP), (strept)avidin–biotin-peroxidase complex (ABC–PO), and immunogold–silver methods may all be suitable, depending on the particular antigen and the condition of the tissue. If it is necessary to unmask the antigen by proteolytic enzyme (for example trypsin) treatment, this can generally be performed as usual, provided this is taken into account when uncovering the target DNA in the second (hybridization) part of the double staining (see below).

Three well-known incubation protocols are given for IPO staining (*Protocols 9–11*).

Protocol 9. Indirect immunoperoxidase staining of paraffin sections

Reagents

- PBS, pH 7.4
- normal (e.g. swine) serum
- 1% bovine serum albumin in PBS (PBS–BSA)
- antibodies (see Steps 5 and 8)

Method

1. Follow *Protocol 7*.[a]

Protocol 9. *Continued*

  2. Rinse the sections by three changes of PBS for 5 min each.

  3. Block the sections by incubating with normal (e.g. swine) serum diluted 1:50 in BSA–PBS.

  4. Carefully blot the slides around the sections.

  5. Apply the first (e.g. rabbit) antibody at a suitable dilution in BSA–PBS.

  6. Incubate for 30–60 min at room temperature.

  7. Wash the sections in three changes of PBS for 10 min each; preferably using a rocker table.

  8. Apply the second, peroxidase-conjugated antibody (e.g. swine–anti-rabbit) in BSA–PBS and incubate for 30 min at room temperature.

  9. Repeat Step 7.

  10. Develop using DAB/H_2O_2 substrate (see *Protocol 12*).

  11. Leave the sections in PBS and continue with DISH.

  a Proteolysis is performed at this stage if appropriate.

When the staining intensity after the indirect IPO protocol (*Protocol 9*) is insufficient, alternative incubation protocols may be used. For this purpose the most reliable are the PAP (*Protocol 10*) and the ABC–PO (*Protocol 11*) methods. Of these, we generally use the PAP method in combination with polyclonal primary antibodies, whilst the ABC–PO method is preferred in combination with monoclonal primary antibodies. However, care should be taken not to stain the sections too intensely to avoid masking the target DNA. This may happen readily when both antigen and target DNA have the same cellular localization, for example both in the nucleus or in the cytoplasm. Theoretically, the use of the ABC–PO method may lead to spurious double staining if biotinylated probes or antibodies are used for hybridization. However, in practice, in the majority of cases these particular reagents are sufficiently masked by the DAB polymer, and/or their binding to the tissue is released by the rather radical (pre)treatment used for the DISH.

Protocols 10 and *11* give the methods for PAP and ABC–PO, in this case for polyclonal and monoclonal primary antibodies, respectively.

After the final immuno-incubation with the peroxidase conjugate, the sections have to be incubated with a substrate solution to achieve a stain. There are several possibilities, but, because treatment with organic solvents has to be performed for DISH, the number is limited to the only insoluble precipitate-generating substrate, namely diaminobenzidine (DAB)–H_2O_2 (*Protocol 12*). Apart from its insolubility, the resulting red-brown pigment has other advantages, such as its fine, delicate localization in the tissue, its nice contrast with certain blue pigments revealed by alkaline phosphatase-based detection systems, and also the possibility to modify and intensify its colour by means of certain heavy metal ions such as nickel and silver (17, 18).

Protocol 10. Peroxidase–anti-peroxidase (PAP) staining for polyclonal antibodies

(In this protocol primary rabbit antibodies are used as an example)

Note: DAB is potentially carcinogenic — see *Protocol 12*, Step 1

Reagents

- PBS
- normal swine serum
- antibody (rabbit)

- unlabelled swine–anti-rabbit immuno-globulin
- (rabbit) PAP complex
- DAB/H_2O_2

Method

1. Follow *Protocol 7.*[a]
2. Rinse the sections in three changes of PBS for 5 min each.
3. Pre-incubate in 10% normal swine serum in PBS for 10–20 min at room temperature.
4. Remove excess serum from the slide.
5. Incubate with specific (rabbit) antibody, optimally diluted in PBS, for 30–60 min at room temperature.
6. Wash in three changes of PBS, for 10 min each.
7. Repeat Step 4.
8. Incubate with (unlabelled) swine–anti-rabbit immunoglobulin, optimally diluted (e.g. 1:5), in PBS, for 30 min at room temperature.
9. Repeat Step 4.
10. Incubate with (rabbit) PAP complex, diluted (e.g. 1:200) in PBS, for 30 min at room temperature.
11. Repeat Step 4.
12. Develop using DAB/H_2O_2 substrate (see *Protocol 12*).
13. Leave the sections in PBS and continue with DISH.

[a] Proteolysis is carried out at this stage if appropriate

Protocol 11. (Strept)avidin–biotin–peroxidase complex (ABC–PO) staining for monoclonal antibodies[a]

Note: DAB is potentially carcinogenic – see *Protocol 12*, Step 1

Reagents

- PBS
- PBS–BSA (*Protocol 9*)
- normal rabbit serum
- antibodies, e.g. mouse monoclonal

- biotinylated rabbit–anti-mouse
- pre-formed (strept) avidin–biotin–peroxidase complex
- DAB/H_2O_2

Protocol 11. *Continued*

Method

1. Follow *Protocol 7.*[b]
2. Rinse the sections in PBS for three changes of 5 min each.
3. Pre-incubate sections with normal rabbit serum, diluted 1:50 with PBS–BSA, for 10–20 min at room temperature.
4. Remove excess serum.
5. Incubate with specific (mouse, monoclonal) antibody, diluted in PBS–BSA, at room temperature for 60 min.
6. Rinse in three changes of PBS for 10 min each.
7. Incubate with biotinylated rabbit–anti-mouse antibody, diluted in PBS–BSA, for 30 min at room temperature.
8. Repeat Step 6.
9. Incubate with pre-formed (strept)avidin–biotin–peroxidase complex, diluted in PBS–BSA (e.g. 1:200) for 60 min at room temperature. [c]
10. Repeat Step 6.
11. Develop the peroxidase with DAB/H_2O_2 substrate (*Protocol 12*).
12. Leave the sections in PBS and continue with DISH.

[a] Protocol based on ref. 16.
[b] Proteolysis can be carried out at this stage if appropriate.
[c] Instead of Step 9, peroxidase-conjugated streptavidin may be used, diluted in PBS–BSA, e.g. 1:500, for 60 min.

Protocol 12. Peroxidase staining with diaminobenzidine (DAB)/ H_2O_2

Note: DAB is potentially carcinogenic (see Step 1)

Reagents

• DAB   • H_2O_2
• PBS

Method

1. Prepare the DAB solution. As DAB is potentially carcinogenic make it up as a 10-fold concentrated solution and store in aliquots at $-20°C$. Weigh and dissolve the substance in a fume cupboard or other safety cabinet, and wear gloves and a dust mask. Stock solution of DAB:
   • Dissolve 5 g diaminobenzidine dihydrochloride in 1 litre PBS (pH 7.4).
   • Stir for 1 h in a dark place.

- Filter the solution 3 times.
- Make 10 ml aliquots and freeze them at − 80°C.
- Store at − 20°C.

  Directly before use: add one 10 ml aliquot of DAB to 100 ml PBS. Add 70 μl H_2O_2 (30% v/v).

2. Add the DAB/H_2O_2 solution to the (wet) slides in a Coplin jar.
3. Stain for 3–5 min in a dark place.
4. Remove the DAB solution carefully and inactivate it with bleaching liquid.
5. Rinse the sections with tap water, followed by distilled water.
6. Leave the sections in PBS.

2.3.2 Immunostaining for frozen sections.

Protocols 9, 10, and *11* can be used for IPO staining of frozen sections, but different pre-treatments are required. In many instances, for example for the demonstration of cell surface antigens, acetone is the fixative of choice but PLP or paraformaldehyde are better for some antigens. On the other hand, the blocking of endogenous peroxidase may lead to problems on frozen sections. If the tissue contains much endogenous peroxidase, the use of immuno-alkaline phosphatase (IAP) or immuno-beta-galactosidase (IBG) methods should be considered. For frozen sections *Protocols 9, 10*, and *11* can be modified as follows:

(a) Follow *Protocol 8*.

(b) Continue from Step 3 of *Protocols 9, 10*, or *11*.

For frozen sections, immuno-alkaline phosphatase (IAP) and -beta-galactosidase (IBG) staining methods for antigen detection may be more effective. Best results can be obtained with the staining of cytoplasmic antigens. For tissue that contains a great deal of endogenous peroxidase, the inactivation of which may be difficult in frozen sections, these alternative methods can be used advantageously. However, as with the IPO methods, many of the possible pigments used for alkaline phosphatase-based techniques are not compatible with the conditions used for subsequent hybridostaining. For this reason only a few substrate–chromogen combinations are possible. In *Protocols 13* and *14* we describe methods which reveal red-coloured pigment that is reasonably resistant to organic solvents.

For frozen sections, it is also possible to use IBG staining (*Protocol 15*), in particular for cytoplasmic antigens. In this way a gentle blue-green precipitate is formed, which offers a nice contrast to the DAB/peroxidase, as well as to the NBT/BCIP alkaline phosphatase pigment. A disadvantage of the method is its rather long incubation time before clear staining is visible.

Protocol 13. Alkaline phosphatase–anti-alkaline phosphatase (APAAP) staining for frozen sections using monoclonal antibodies [a]

Reagents

- PBS
- normal rabbit serum
- PBS–BSA (see *Protocol 9*)
- unlabelled rabbit–anti-mouse immunoglobulin
- APAAP

Method

1. Pre-incubate with normal rabbit serum, diluted 1:50 in PBS–BSA, for 15 min.

2. Carefully blot the slides around the sections.

3. Incubate with monoclonal antibody, at the appropriate dilution in PBS–BSA, for 60 min at room temperature.

4. Wash in three changes of PBS for 10 min each.

5. Incubate with unlabelled rabbit–anti-mouse immunoglobulin, diluted in PBS–BSA (e.g. 1:25 [b]), for 30 min at room temperature.

6. Wash in three changes of PBS for 10 min each.

7. Incubate with APAAP (mouse), diluted (e.g. 1:50 [b]) in PBS–BSA, for 30 min at room temperature.

8. Wash in three changes of PBS for 10 min each.

If, after a pilot experiment the staining is too weak, Steps 5–8 may be repeated with 15 min incubations.

9. Develop alkaline phosphatase staining using *Protocol 14.*

[a] Protocol based on ref. 19.
[b] In this protocol the dilutions are based on our experience with antibodies purchased from DAKO. They may be adapted according to one's own experience and antibodies.

Protocol 14. Alkaline phosphatase staining with Naphthol ASBI phosphate and diazotized New Fuchsin

Reagents

- New Fuchsin
- 2 M HCl
- sodium nitrite
- 0.05 M Tris–HCl, pH 8.7
- levamisole
- Naphthol ASBI phosphate
- dimethylformamide
- PBS

Method

1. Prepare solution A by mixing:
   - 5% (v/v) New Fuchsin in 2 M HCl   20 μl
   - 4% (w/v) freshly prepared sodium nitrite   50 μl
   - Mix well and wait 1 min, then add:
   - 0.05 M Tris–HCl buffer, pH 8.7   10 ml
   - levamisole [a]   2.5 mg

2. Prepare solution B by dissolving 5 mg Naphthol ASBI phosphate in 60 μl dimethylformamide in a glass tube.

3. Add solution A to solution B.

4. Mix well and filter the solution directly on to the slide.

5. Incubate for 30–45 min in a closed and moistened box at room temperature.

6. Wash the sections in running tap water and then in PBS.

7. Leave the sections in PBS until DISH is performed.

NB Result: positive cells are stained red.

[a] When no endogenous alkaline phosphatase is expected, this may be omitted.

Protocol 15. Indirect immuno-beta galactosidase staining for monoclonal antibodies on frozen sections [a]

Reagents

- PBS–BSA (see *Protocol 9*)
- sheep–anti-mouse–beta-galactosidase
- solution A [b]: dissolve 54 mg 5-bromo-4-chloro-3-indolyl-beta-galactopyranoside (BCIG; X-GAL) in 1 ml dimethylformamide in a glass tube.
- solution B [b]: dissolve 2 mg $MgCl_2 \cdot 6H_2O$, 10 mg potassium ferricyanide, and 13 mg potassium ferrocyanide in 10 ml PBS.

Method

1. Follow Steps 1–4 of *Protocol 3*.

2. Incubate with sheep–anti-mouse–beta-galactosidase, diluted 1:10 in PBS–BSA, for 30 min at room temperature.

3. Wash in three changes of PBS for 10 min each.

4. Prepare beta-galactosidase substrate solution by mixing 50 μl solution A with 5 ml solution B.

5. Drip the substrate solution on to the slides and stain for 60 min at 37°C.

Protocol 15. *Continued*

6. Wash in tap water, leave slides in PBS, and continue with DISH.

NB Result: positive cells are stained blue-green.

---

[a] Protocol based on ref. 20.
[b] Solutions A and B may be prepared in advance and stored in aliquots, at −20 and 4°C, respectively.

2.4 DNA *in situ* hybridization

After the immunostaining is complete DISH can be performed, provided that the sections have been stored in the appropriate way (*Protocol 15*).

2.4.1 Probes for DISH

Double-stranded DNA probes can be prepared by nick-translation or random primer labelling in the presence of a labelled deoxynucleotide. For detailed protocols see Chapter 1 of this volume. However, nowadays in many instances DNA probes may be obtained commercially, either as part of complete hybridization kits, or separately. Optimal signals can be obtained using probes of about 50–200 base-pairs long. Insert probes should be used as much as possible because some plasmids, for example pBR322, occasionally seem to hybridize with genomic DNA, resulting in non-specific staining which should be avoided as much as possible in double-staining experiments.

The composition of an optimal hybridization mixture, in which the probe should be dissolved to a concentration of 1–2 ng/μl, is given in *Table 1*.

2.4.2 Choice of probe label

Although several alternatives are available the labels currently used most for non-radioactive DNA probes are biotin (21) and digoxigenin (developed by Boehringer–Mannheim). Our experience with DISH–IC double-staining methods are mainly based on the use of such probes. The choice of the label to be used is rather arbitrary with respect to sensitivity. Despite some reports which favour the use of digoxigenin-labelled probes in this respect, our experience, and that of others, is that there is hardly any difference between

---

Table 1. Composition of hybridization mixture for DNA *in situ* hybridization

- 50% formamide
- 10% Denhardt solution (0.2% Ficoll, 0.2% BSA, 0.2% polyvinylpyrrolidone)
- 2 × SSC: 0.3 M sodium chloride, 3 mM sodium citrate, pH 6.8
- 10% dextran sulfate
- 250 μg/ml denatured sonicated salmon sperm DNA

---

these and biotin-labelled probes. However, if a choice has to be made some points have to be kept in mind. Biotin-labelled DNA probes, compared with digoxigenin-labelled probes, have the advantage that they are more universally available. This is also the case for the reagents that have to be used for the preparation of these probes and the antibodies or reagents that are needed for their detection. However, another consideration is the presence of endogenous biotin in the tissue. Because certain tissues, like liver or brain, contain an amount of biotin which can be detected easily immunohistochemically, the use of biotinylated probes should be avoided for such tissue. Digoxigenin is not an endogenous substance, and its use as a probe label often yields cleaner sections.

An advantage of digoxigenin-labelled probes is that spurious double staining, owing to the interference of ABC protocols used for immunostaining, can be avoided. However, the extended detection protocols, used by us, for digoxigenin-labelled probes make use of streptavidin and biotin-based incubation sequences. On the other hand, the risk of spurious double staining due to interference of two biotin/avidin based detection systems is mainly theoretical as explained in Section 2.3.1.

2.4.3 Permeabilization of tissue

One of the most important steps in DISH protocols is the permeabilization of the tissue by proteolytic enzyme treatment, for example by proteinase K or pepsin. It is an empirical procedure: concentration and incubation time should be determined in advance by pilot experiment. A minor variation in concentration may lead to incomplete uncovering of target DNA, or on the other hand impairment of tissue morphology. For IC–DISH double-staining, trypsin pre-treatment for antigen detection should be taken into account: the proteinase K treatment before DISH should be modified or even deleted.

Frozen and paraffin sections also require different pre-treatments. For paraffin sections digestion with a proteolytic enzyme (mostly proteinase K or pepsin/HCl) is always necessary, whereas this is not an absolute pre-requisite for frozen sections. The concentration of the enzyme and/or the time of digestion are considerably lower, and shorter, respectively, for frozen sections.

2.4.4 Avoidance of background staining

Apart from background staining due to adhesion of the probe to certain tissue elements, adherence of the probe to the pre-formed DAB pigment may occasionally be seen in double-staining experiments, resulting in spurious double staining. This can often be avoided by acetylation of the tissue with acetic anhydride before the hybridization itself is started (4, 22) (*Protocol 16*).

2.4.5 *In situ* hybridization protocols for DNA

We first give a general protocol for DISH that can be used for biotin- as well as digoxigenin-labelled probes. Separate protocols are given for paraffin and frozen sections followed by hybridocytochemical detection protocols. For the DISH protocols it is assumed that the sections are already immunostained.

i. DISH on paraffin sections

Protocol 16. DISH on paraffin-embedded sections[a]

Reagents
- PBS
- 0.05 M Tris–HCl, pH 7.6, 5 mM EDTA
- proteinase K
- hybridization mixture (see *Table 1*)
- 2 × SSC: 0.3 M sodium chloride, 30 mM sodium citrate
- 0.25% acetic anhydride in 0.1 M triethanolamine, 0.09% NaCl
- alcohol series
- 0.1 × SSC, 50% formamide

Method
1. Wipe off excess PBS from the wet slide.
2. Digest the sections with proteinase K, dissolved at a concentration of, e.g. 0.5 mg/ml in 0.05 M Tris–HCl buffer, pH 7.6, 5 mM EDTA. Drip the solution on the slides and incubate for 15–30 min at 37°C in a pre-heated and humidified box.[b,c]
3. Wash in three changes of distilled water for 5 min each.
4. Pre-hybridize with hybridization mixture (Section 2.4.1) by dripping 10–20 µl on to the section, covering it with a coverslip of appropriate size, and incubating in a humidified box for 60 min at 37°C.[d]
5. Remove the coverslips by rinsing in 2 × SSC.
6. Wash in PBS.
7. Acetylate the sections by incubating in 0.25% acetic anhydride for 10 min.[e]
8. Wash in double-distilled water.
9. Dehydrate the sections in a graded series of ethanol (50, 70, 90, 100%). The ethanol should be of the highest quality. Use distilled water for dilution.
10. Air-dry the sections for about 10 min.
11. Drip 10–20 µl (1–2 ng/µl) biotinylated or digoxigenin-labelled probe in hybridization mixture on to the sections.
12. Coverslip the sections immediately.
13. Put the slides on a heating plate, with a temperature indicator, at a temperature of 95–100°C and denaturate the probe and target DNA together for 7 min.
14. Transfer the slides to a humidified box, close the box, and incubate in an oven at 37°C overnight.

---

15. Remove the coverslips by placing the slides in pre-warmed 2 × SSC solution at 37°C.

16. Wash stringently in three changes of 0.1 × SSC, containing 50% formamide, for 20 min each at 37°C.

17. Rinse in three changes of PBS for 5 min each and proceed to *Protocol 17*.

[a] Protocol based on refs 7, 21, and 23.
[b] The proteinase K concentration should be determined empirically in advance.
[c] Alternatively, 0.25% pepsin in 0.2 M HCl can be used for 10–30 min (24).
[d] Steps 4–6 are optional, depending on the presence of background staining.
[e] This step is optional, and should only be used if spurious double staining is a problem.

ii. *DISH on frozen sections*

For frozen sections a similar protocol can be used by amending *Protocol 16* as follows:

(a) Remove excess PBS from the slide.

(b) Follow Steps 4–17 from *Protocol 16*.

2.4.6 Detection of the hybridized probe

For double IC–DISH experiments, simple one-step incubation protocols are often not sensitive enough. Therefore, the hybrid-bound biotin should be treated as a hapten. In this way the same protocols can be used for the detection of biotin and digoxigenin apart from the primary antibody.

Protocol 17. Detection of biotin- or digoxigenin-labelled DNA hybrid, using alkaline phosphatase and NBT/BCIP

Reagents

- antibodies: monoclonal mouse-anti-biotin (Boehringer) or monoclonal mouse-anti-digoxin (Sigma) (or mouse-anti-digoxigenin, Boehringer),
- biotinylated horse–anti-mouse immunoglobulin
- streptavidin
- biotinylated alkaline phosphatase (BRL, DAKO)
- 1% blocking reagent (Boehringer–Mannheim) in PBS (BR–PBS)
- PBS
- Tween-20
- alkaline phosphatase buffer (AP): 0.1 M Tris–HCl, pH 9.5, 0.1 M NaCl, 50 mM $MgCl_2$
- Nitroblue Tetrazolium (NBT) solution: 75 mg/ml in 70% dimethylformamide
- 5-bromo-4-chloro-3-indolyl phosphate (BCIP) solution: 50 mg/ml in 70% dimethylformamide
- levamisole

Method

1. From *Protocol 16* wipe off excess PBS from the wet slide.

2. Pre-incubate the sections in BR–PBS for 30 min.

Protocol 17. *Continued*

3. Remove excess liquid around the section.

4. (a) *For biotin-labelled hybrids*: incubate with monoclonal mouse-anti-biotin, diluted 1:250 in BR–PBS, for 30 min.

   (b) *For digoxigenin-labelled hybrids*: incubate with monoclonal mouse-anti-digoxin (diluted 1:250 in BR–PBS) or mouse-anti-digoxigenin (diluted 1:100 in BR–PBS) for 60 min.

5. Wash with three changes of PBS + 0.1% Tween-20 for 10 min each.

6. Incubate with biotinylated horse–anti-mouse Ig (1:250 in BR–PBS) for 30 min.

7. Repeat Step 5.

8. Incubate with streptavidin, 1:2000 in BR–PBS, for 30 min.

9. Repeat Step 5.

10. Incubate with biotinylated alkaline phosphatase (1:2000 in BR–PBS) for 30 min.

11. Repeat Step 5.

12. Rinse once in AP buffer for 3 min.

13. Develop alkaline phosphatase (25) with a mixture consisting of:

    • AP buffer   5 μl
    • NBT solution   22 μl
    • BCIP solution   17 μl

    If endogenous alkaline phosphatase is expected, add 1.25 mg levamisole per 5 ml substrate solution. Incubate for 30–60 min (or longer), monitoring staining intensity microscopically.

14. Stop the reaction by washing in three changes of water for 3 min each.

15. Mount the slides in an aqueous mountant, e.g. Kaiser's glycerin–gelatin.

NB Result: positive cells contain purple-blue dots.

Protocol 18. Detection of biotin- or digoxigenin-labelled DNA hybrids using peroxidase/DAB

Note: DAB is potentially carcinogenic — see *Protocol 12*, Step 1

Reagents

• Streptavidin–biotin–peroxidase complex (ABC–HRP) (DAKO)
• PBS/0.1% Triton-X-100
• DAB/H_2O_2 (*Protocol 12*)

Method

1. Follow Steps 1–7 from *Protocol 17.*

2. Incubate with pre-formed streptavidin–biotin–peroxidase complex[a] (ABC–HRP) in BR–PBS for 30 min. [b]

3. Wash with three changes of PBS containing 0.1% Triton-X-100 for 10 min each.

4. Develop peroxidase with DAB/H_2O_2 according to *Protocol 12* or *20.*

5. Mount the sections with aqueous mountant (*Protocol 17*).

[a] This should be prepared according to the manufacturer's instructions.
[b] An alternative to Step 2 is to incubate with streptavidin–HRP (Detek-HRP, Enzo), diluted 1:250 in BR–PBS, for 30 min. Step 2 occasionally gives a faint background staining which is diminished when this alternative is used. However, the staining intensity is also slightly reduced.

2.4.7 Alternative peroxidase staining

For *Protocol 18*, DAB can be combined with nickel (Ni) alone or with Ni and silver to enhance staining compared with DAB alone. Use of DAB/Ni reveals blue/grey to dark-blue staining instead of brown/red. With silver enhancement, black granular staining develops. Peroxidase staining using DAB/Ni and silver is recommended for tissue with little target DNA and can be used in combination with common IPO antigen detection with DAB.

The use of DAB/Ni without silver enhancement is sometimes useful if the target DNA is abundantly present, and contrasts nicely with the common brown-red DAB pigment formed in the immunoperoxidase antigen detection.

In *Protocol 19* we give the modification of the peroxidase staining, using DAB/nickel as substrate, and subsequently in *Protocol 20* the silver enhancement procedure.

Protocol 19. Peroxidase staining with DAB/nickel/H_2O_2 [a]

Note: DAB is potentially carcinogenic — see *Protocol 12*, Step 1

Reagents

- aliquots of 10 mg DAB in 1 ml 50 mM Tris– HCl buffer, pH 7.6; store at −20°C
- 50 mM Tris–HCl buffer, pH 7.6, containing 0.7 g/l $NiCl_2.6H_2O$ (this solution may be stored at 10-fold concentration)
- 30% (v/v) H_2O_2
- 4% paraformaldehyde
- Depex

Method

1. Before use:
   - thaw 1 ml DAB/Tris solution;
   - add 50 ml Tris–HCl buffer containing nickel chloride;
   - add 3.5 μl 30% H_2O_2.

Protocol 19. *Continued*

2. Put the slides in a Coplin jar, and add the DAB/nickel/H$_2$O$_2$ solution.

3. Stain for 5 min in a dark place.

4. Remove the solution from the jar and inactivate it with bleaching liquid.

5. Rinse the sections thoroughly with tap water, followed by distilled water.[b]

6. Fix the sections with 4% paraformaldehyde for 5 min.

7. Wash sections with water.

8. Dehydrate sections and mount with either Depex or aqueous mounting medium, depending on the nature of the pigment used in the immunostaining.

[a] Protocol based on refs 17, 18, 26.
[b] If the silver intensification (*Protocol 20*) is to be added, proceed from this point without aldehyde treatment.

Protocol 20. Silver intensification of DAB/nickel staining [a]

Reagents

- 1% sodium acetate
- 1% acetic acid
- graded ethanol series
- xylene
- Depex
- *solution A*[b]: dissolve 80 g sodium acetate in 500 ml double-distilled water, add 5.6 ml glacial acetic acid and 100 ml 1% silver nitrate. Add 10 ml 1% cetylpyridinium chloride under vigorous stirring, keep the solution overnight at 4°C and filter. Add 60 ml 1% Triton-X-100 and make up the solution to 800 ml with distilled water
- *solution B*[b]: dissolve 5 g sodium tungstate in 100 ml double-distilled water
- *solution C*[b]: dissolve 0.2 g ascorbic acid in 100 ml double-distilled water

Method

1. Use sections stained with DAB/nickel according to *Protocol 19*.

2. Wash the sections thoroughly with distilled water.

3. Rinse sections in 1% sodium acetate in double-distilled water for 10 min.

4. Prepare the silver developing solution by slowly adding 1 ml of solution B and 1 ml solution C to 8 ml of solution A and mix thoroughly. Filter this solution (for example through a 0.45 μm Millipore filter) and immediately drip on to the sections.

5. Incubate for 0.5–8 min in a closed (dark) box at room temperature. Monitor staining intensity after 15 sec and repeat this after 30 and 60 sec, and so on until the desired staining intensity is reached. **Do not** incubate longer than the moment that clearly visible staining can be observed because background staining will rapidly appear.

6. Wash the sections in 1% acetic acid for 2–10 min. This will prevent further reaction.

7. Wash in 1% sodium acetate for 1–2 min.

8. Wash in tap water.

9. Dehydrate the sections in a graded ethanol series and xylene.

10. Mount sections with Depex, etc.

---

[a] Protocol according to refs 18, 26.
[b] These solutions may be prepared in advance. Prepare aliquots of 8, 1, and 1 ml of solutions A, B, and C, respectively, and store at −20 °C.

2.4.8 Which combinations of IC and DISH protocols can be used?

In all cases the individual immunohistochemical and hybridocytochemical reactions should be performed first individually. Double staining is only meaningful when both single procedures show clear, non-diffuse staining with no background staining. If the immunostaining is very heavy it is likely that it will be too dominant after double staining, and the pigment may mask the target DNA or its labelled hybrid. In such a case the immunostaining should be modified so that the staining intensity is more or less of the same strength as that of the expected hybridostaining. This may be achieved in several ways, for example by cutting the incubation time of the chromogenic enzyme reaction, by using a higher dilution of the primary antibody, or by using a more simple staining protocol (for example an indirect instead of an ABC method).

i. Paraffin sections

Paraffin sections can only be immunostained with one of the peroxidase protocols (*Protocols 9, 10,* or *11*). If there is sufficient target DNA available, *Protocol 17* (NBT/BCIP) can be used for DISH. When the target DNA is less abundantly present, *Protocol 19* (DAB/Ni) (*Plate 13*), or better still *Protocol 20* (DAB/Ni/Ag) (27) may be used. Although these protocols use peroxidase as enzyme label for both immunostaining and DISH, we have never found spurious staining, either due to the presence of the first peroxidase labels, or due to colour conversion of the first (immunostaining) DAB pigment by the modified DAB substrates with nickel and silver.

ii. Frozen sections and cell monolayers

For frozen sections the first choice for immunostaining is also one of the peroxidase protocols (*Protocols 9, 10,* or *11*). The resulting red-brown DAB polymer contrasts nicely with the NBT/BCIP reaction product, which is revealed after DISH with alkaline phosphatase (*Protocol 17*) (*Plate 11a*). This method is also suitable for cells grown on coverslips (*Plate 12*).

For frozen sections the IAP protocol using Naphthol ASBI phosphate and New Fuchsin as substrate may be also be used (*Protocols 13* and *14*). These protocols are preferred when there is much endogenous peroxidase present. For tissue immunostained in this way at least two DISH protocols can be used. Combination with *Protocol 17* (NBT/BCIP) (*Plate 11b*) or *Protocol 19* (DAB/Ni) is recommended. In combination with *Protocol 17*, two alkaline phosphatase-based methods are used with the theoretical risk of spurious double staining. However, we have never found such spurious staining, probably due to the destruction of the APAAP complex during the DISH protocol. As an alternative, *Protocol 19* may be used, giving a dark-blue staining which also contrasts nicely with the red immunostaining.

Another possibility for frozen sections that contain endogenous peroxidase is the combination of *Protocol 15*, with beta-galactosidase as enzyme label, and *Protocol 17* (NBT/BCIP) (*Plate 11c*). Because the blue-green immunostaining is rather diffuse, it is less suitable for staining of cell membrane antigens, but appropriate for cytoplasmic antigens.

iii. Summary

For frozen sections the first choice is a combination of *Protocol 9, 10,* or *11* (IPO, DAB, brown) with *Protocol 17* (NBT/BCIP, purple/blue). The second choice is the combination of *Protocols 13* and *14* (IAP, red) with *Protocol 17* (NBT/BCIP) and then with *Protocol 19* (DAB/Ni). The third option is the combination of *Protocol 15* (IBG, green) with *Protocol 17* (NBT/BCIP) and then with *Protocol 19* (DAB/Ni).

For cytological preparations or cells grown on coverslips the first choice is also a combination of *Protocol 9, 10* or *11* with *Protocol 17*.

For paraffin sections one of *Protocol 9, 10,* or *11* (IPO, DAB), is best combined with *Protocol 17* (NBT/BCIP) for DISH. If hybridostaining does not, or almost not, reveal any signal, *Protocol 20* (DAB/Ni/Ag) is recommended.

3. Combination of non-radioactive RNA *in situ* hybridization with immunocytochemistry

3.1 Introduction

Although the use of DISH on tissue sections finds its main application in the demonstration of viral DNA, RNA *in situ* hybridization (RISH) is more universally applicable principally because the expression of all genes, either endogenous or of microbial origin, can be studied at the messenger RNA level. Until now the technique has been practised predominantly with radioactive probes, followed by autoradiography. The non-radioactive procedures, described in the literature, are relatively scarce and their main application is reserved for tissues in which the genes of interest are abundantly expressed.

The combination of RISH with IC theoretically allows the detection of any pair of RNA and tissue antigens for which specific reagents are available. In this way the transcriptional versus the translational regulation of the expression of a given gene can be studied, as can the effect of the transcription of a gene on the synthesis of a protein encoded by another gene, or vice versa (27). Sometimes, if certain (for example viral) genes are abundantly expressed, RISH can be more useful than DISH to identify infected cells. The technique can be combined with IC, using antibodies to cell-specific antigens in order to characterize the host cells. In fact such a procedure (using radioactive probes) was described earlier than the combination of DISH and IC (4).

Here we will consider the practical application of the combination of both techniques. Double staining of RISH with IC will have some of the same problems as the combination of IC with DISH, but other factors have also to be taken into account. The procedure for tissue pre-treatment to uncover the target nucleic acid is more aggressive for DISH than for RISH. Moreover, a DNA target has to be denatured at much higher temperatures than a RNA target, so the chance that vulnerable tissue epitopes will stay intact is higher after RISH than after DISH, although there is still some risk of impairment. For this reason most authors, who practised radioactive *in situ* hybridization as part of a combination staining, first carried out immunostaining and then RISH (4–6). However, the main problem we encounter in RISH is how to keep the RNA unimpaired during the tissue pre-treatment, for example, during a preceding immunostaining. In particular, when hyper-immune, polyclonal antisera are used for one of the immuno-incubations, the risk of RNA degradation by the RNase present in the immune serum is considerable. Therefore, a number of precautions have to be taken if immunostaining precedes the RISH in order to protect the RNA.

As stated above, the RISH procedure is rather mild as compared with DISH, and, therefore, the risk that tissue antigens will be affected is much less. For this reason, and because of the vulnerability of the RNA to ubiquitous RNases, we prefer a staining sequence in which the *in situ* hybridization is performed first. It must be kept in mind, however, that this sequence will hardly work for very sensitive antigens. Here we will deal with protocols that are suitable for antigens that are stable enough to withstand routine formalin fixation and paraffin embedding and, in general, also withstand the RNA hybridization protocol.

Because many of the procedures to be used for the IC–RISH combination are the same as for the combination of IC–DISH, only the differences will be given here. We will describe here some of the protocols that have been succesfully used by us.

3.2 Precautions

To be RNase-free glassware should be extensively washed with a suitable product (Decon, Lipsol), rinsed with distilled water, and dried overnight at 180°C.

Only RNase-free water and solutions should be used: the water should be double distilled and treated with DEPC.

3.3 Preparation of microscope slides for paraffin sections

Use *Protocol 1*: be particularly careful to use pure distilled water, including for making up the glutaraldehyde. Wear gloves when the slides are handled after their preparation. When sticking the 4–6 μm sections on the APES-coated slides, use double-distilled water and avoid any contact of the slides with bare hands. Then keep the slides in clean, dust-free boxes. Do not make too many slides in advance because the quality of the RNA may be impaired.

3.4 Preparation of specimens

The corresponding DISH protocols can be used. Our experience with RISH is predominantly based upon the use of routinely, buffered formalin-fixed and paraffin-embedded tissue. For cytological preparations paraformaldehyde (*Protocol 4*) or acetic acid–paraformaldehyde (*Protocol 6*) can be used.

3.5 Probes for RISH

For RISH cDNA-, RNA- and oligonucleotide probes can be used. We will describe protocols which are suitable for biotin- or digoxigenin-labelled RNA and oligonucleotide probes.

3.6 RISH

The methods used for RISH are almost as many as there are authors. They are dependent on the nature of the probe, the required stringency of the hybridization, and particularly one's own experience.

Protocol 21 describes a method for RISH which is used by us for different probes and tissues. It is suitable for the demonstration of cytoplasmic or nuclear RNA in paraffin-embedded tissue and can be used for biotin- and digoxigenin-labelled RNA probes.

Protocol 21. RISH with biotinylated or digoxigenin-labelled probes[a]

Reagents

Note: All **materials** should be DEPC-treated and gloves should be worn

- DEPC-treated water
- 0.2 M HCl
- 2 × SSC: 0.3 M sodium chloride, 30 mM sodium citrate
- 0.05 M Tris–HCl, pH 7.6
- proteinase solution: 10 mg/ml in 0.05 M Tris–HCl, pH 7.6
- 0.4% paraformaldehyde in PBS (*Protocol 4*)
- 0.25% acetic anhydride
- 0.1 M triethanolamine

- graded alcohol series
- deionized formamide containing 50 μg/ml tRNA
- 5 × SSC + 25% dextran sulfate in water
- probe, dissolved in formamide
- 0.1% Triton-X-100
- 0.1 × SSC

1. Follow *Protocol 7.*

A. *Pre-treatment*

1. Wash in water for 5 min.
2. Incubate the sections in 0.2 M HCl for 20 min, then rinse in water.
3. Incubate in 2 × SSC for 10 min at 70°C, then rinse in water.
4. Rinse in 0.05 M Tris–HCl buffer, pH 7.6, for 1 min.
5. Place the wet sections in a pre-warmed incubation box and cover them with 0.5 ml of freshly prepared proteinase K solution. Incubate for 60 min at 37°C.
6. Stop the reaction by immersing in pre-cooled water (4°C) for two changes of 10 min each.
7. Optional DNase or RNase treatment may be performed for 1 h at 37°C followed by washing with three changes of water for 2 min each. [b]
8. Immerse in 0.4% paraformaldehyde in PBS for 20 min.
9. Rinse in water.
10. Acetylate the sections with 0.25% acetic anhydride in 0.1 M triethanolamine for 10 min, then rinse in double-distilled water. Use three changes of 2 min each. [c]
11. Dehydrate the sections in graded ethanol series (70, 90, 98%) and air-dry.

B. *Hybridization*

1. Prepare the hybridization mixture as follows in a 1.5 ml reaction tube:
   - deionized formamide containing 50 μg/ml tRNA   to 100 μl
   - 5 × SSC [d] + 25% dextran sulfate in water   40 μl
   - water   10 μl
   - probe, dissolved in formamide [e]   x μl

   mix well.
2. Apply 20–40 μl of hybridization mixture, depending on the size of the section.
3. Cover the sections with a coverslip (*Protocol 3*) of appropriate size.
4. Denature the sections on a hot plate (e.g. Shandon) or in an oven, with temperature control, at 60–65°C for 7 min.
5. Place slides in a tightly closed and moistened incubation chamber.
6. Hybridize overnight at 37–70°C. [d]

Protocol 21. *Continued*

C. *Post-hybridization*

1. Immerse the slides in 2 × SSC to loosen the coverslips.
2. Incubate the slides with two changes of 2 × SSC at 37–70 °C[e] for 10 min each.
3. Incubate the slides in 0.1 × SSC at 37–70 °C for 10 min.
4. Incubate in the PBS, containing 0.1% Triton X-100 for 10 min at room temperature.
5. Place the sections in an incubation chamber and continue with one of the detection *Protocols 22* or *23*.

[a] Protocol modified from ref. 28.

[b] This step may be performed in order to control the nature of the obtained signal (RNA or DNA):
  - DNase: mix 5 units RQ-DNase with 50 µl transcription buffer.
  - RNase: dissolve 1 mg/ml RNase A in 0.01 Tris–HCl, pH 8.0, with 1 mM EDTA and 0.5 M NaCl.

[c] To prevent or to diminish possible background staining (22).

[d] The SSC concentration of the hybridization mixture, the hybridization temperature, and the temperature of the post-hybridization washings are dependent on the nature of the probe and the required stringency. For some probes we use, e.g. 2 × SSC, 37 °C, and 37 °C, respectively, but for other (RNA) probes we use, e.g. 0.5 × SSC, 60 °C and 75 °C, respectively.

[e] The concentration of the probe should be determined in advance, with the signals obtained on a Northern blot as a guide. See Chapter 6.

Protocol 22. Detection of biotin- or digoxigenin-labelled RNA hybrids with alkaline phosphatase and NBT/BCIP in combination with IPO (DAB) immunostaining

Reagents

- normal horse serum
- 1% blocking reagent (Boehringer) in PBS (BR–PBS)
- 0.1% Tween-20
- antibodies: (a) monoclonal mouse-anti-biotin (Boehringer–Mannheim) or mono-clonal mouse-anti-digoxin (Sigma) (mouse-anti-digoxigenin, Boehringer–Mannheim), (b) biotinylated horse–anti-mouse (Vector)
- streptavidin (BRL, DAKO)
- biotinylated-AP

Method

1. Pre-incubate with normal horse serum, diluted 1:50 in BR–PBS, for 20 min.
2. Wipe off excess blocking medium.
3. (a) *For biotin-labelled hybrids*: incubate with monoclonal mouse-anti-biotin, diluted 1:250 in BR–PBS, for 60 min.
   (b) *For digoxigenin-labelled hybrids*: incubate with monoclonal mouse-anti-digoxin (1:250) or mouse-anti-digoxigenin (1:100), in BR–PBS, for 60 min.

4. Wash in three changes of PBS/0.1% Tween-20 for 10 min each.

5. Incubate with biotinylated horse–anti-mouse (1:500), in BR–PBS for 30 min.

6. Repeat Step 4.

7. Incubate with streptavidin (1:2000, BRL; 1:100, DAKO), in BR–PBS, for 30 min.

8. Repeat Step 4.

9. Incubate with biotinylated alkaline phosphatase (diluted 1:2000, BRL; or 1:100, DAKO) in BR–PBS for 30 min.

10. Repeat Step 4.

11. Follow steps 12–14 of *Protocol 17*.

12. Keep the sections in PBS at 4°C overnight or continue with immuno-staining.

13. Immunostain the sections with one of the immunoperoxidase pro-tocols (*Protocols 9, 10* or *11* with 12).

14. Mount the sections with aqueous mounting medium.

NB Result: brown-red immunostaining with purple-blue stained RNA.

Protocol 22 is useful in those cases where the antigen and the RNA are spatially separated, for example both are present in different cells or the antigen is present in the cytoplasm and the RNA in the nucleus, or vice versa (*Plate 14b*).

When the RNA and antigens to be detected are localized within the same part of the cell the combination in *Protocol 22* is less suitable. Additionally, if the antigen is part of the cell membrane with the RNA in the cytoplasm, this combination has the disadvantage that the NBT/BCIP reaction product is diffuse so that a certain interference of the pigments can occur, producing a dark-grey colour. In such cases it is preferable to use *Protocol 23*, which results in a combination of black, granular RNA staining with brown-red stained antigen (*Plates 14a* and *15*). Alternatively, *Protocol 24* may be used for tissue containing abundant RNA.

Protocol 23. Detection of biotin- or digoxigenin-labelled RNA hybrids with peroxidase and DAB/Ni with Ag intensification in combination with IPO (DAB) immunostaining

Reagents

- pre-formed streptavidin–biotin–peroxidase complex (ABC–HRP (DAKO)), see *Protocol 18*, Step 2
- 0.1% Tween-20
- 4% paraformaldehyde
- methanol + 0.3% H_2O_2
- DAB/H_2O_2 (*Protocol 12*)

Protocol 23. *Continued*

Method

1. Follow Steps 1–6 of *Protocol 22*.
2. Incubate with pre-formed ABC–HRP complex, in BR–PBS, for 30 min.
3. Wash in three changes of PBS containing 0.1% Tween-20, for 10 min each.
4. Follow Steps 1–5 of *Protocol 19*.
5. Fix the sections with 4% paraformaldehyde for 5 min at room temperature.[a]
6. Wash the sections with water.
7. Incubate the sections in methanol containing 0.3% H_2O_2 for 20 min.[b]
8. Rinse the sections in water, then in PBS.
9. Keep the sections in PBS at 4°C overnight, or continue immediately.
10. Immunostain the sections with peroxidase/DAB according to *Protocol 9*, *10*, or *11*, then *12*.
11. Follow *Protocol 20* for silver intensification of the DAB/Ni staining.[c]
12. Counterstain the sections briefly with Mayer's Haematoxylin, Methyl Green, or Pyronin (optional).
13. Dehydrate the sections and mount with Depex.

NB Result: brown-red immunostaining with black granular staining of RNA and faint blue, green, or red nuclei.

[a] This step is necessary because the DAB/Ni complex will dissociate if the sections are kept overnight in PBS. When the staining sequence is continued immediately, this step is optional.
[b] To block remaining peroxidase activity from Step 2.
[c] Silver intensification for RNA staining with DAB/Ni has to be performed after immunostaining with DAB.

Protocol 24. Detection of biotin- or digoxigenin-labelled RNA hybrids with immunogold-silver detection in combination with IAP immunostaining[a]

Reagents

- 0.2% BSA-C (Aurion) in PBS (BSA–PBS)
- Tween-20
- streptavidin–Au
- 2% glutaraldehyde in PBS
- silver enhancement kit (Amersham, Aurion)

Method

1. Follow Steps 1–4 of *Protocol 21* until step C5.
2. Continue with Steps 1–6 from *Protocol 22*.

3. Incubate the sections with 0.2% BSA-C in PBS containing 0.1% Tween-20 (BSA–PBS), for 5 min.

4. Remove excess liquid from around the sections.

5. Incubate with streptavidin–Au (5 nm globules), in variable dilutions, e.g. 1:200, in BSA–PBS, for 60 min.

6. Wash with three changes of BSA–PBS, for 10 min each.

7. Fix the sections for 10 min at room temperature with 2% glutaraldehyde in PBS.

8. Wash with three changes of PBS for 5 min each.

9. Follow steps 1–8 of *Protocol 13*.[b]

10. Follow *Protocol 14* until Step 7.

11. Wash with several changes of double-distilled (chloride-free) water.

12. Prepare silver enhancement solution[c]:
   - Mix equal aliquots of both solutions in the kit, e.g. 100 μl of each.
   - Let the mixture acclimatize at room temperature for 5–10 min in the dark.

13. Drip the silver solution on to the sections.

14. Incubate in a dark box for 6 min.

15. Wash with three changes of chloride-free water.

16. Repeat Steps 12–15 twice, controlling the signal intensity microscopically.

17. Incubate for a further 1 min in silver solution and wash.

18. Counterstain the section briefly with Mayer's Haematoxilin (optional) and mount the section with an aqueous mounting medium.

Note Result: red cytoplasmic staining of the desired antigen with black silver granules that indicate the RNA as well as faint-blue nuclear staining.

[a] Protocol based on modification of double immunostaining method of ref. 29 with RISH protocol of ref. 11.

[b] The alkaline phosphatase staining should not be too intense. For that reason, in general, a single incubation with APAAP is sufficient.

[c] Follow the instructions of the manufacturer, e.g. Amersham or Aurion. Avoid contact of the sections with metals during these stages.

References

1. Haase, A. T. (1986). *J. Histochem. Cytochem.*, **34**, 27.
2. Grody, W. W., Cheng L., and Lewin, K. J. (1987). *Human. Pathol.*, **18**, 533.
3. Syrjänen, S. M. (1992). In *Diagnostic molecular pathology: a practical approach* (ed. C. S. Herrington and J. O'D. McGee), Vol. 1, pp. 103–39. IRL Press, Oxford.

4. Brahic, M., Haase, A. T., and Cash, E. (1984). *Proc. Natl Acad. Sci. USA*, **81**, 5445.
5. Gendelman, H. E., Moenck, T. R., Narayan O., Griffin, D. E., and Clements, J. E. (1985). *J. Virol. Meth.*, **11**, 93.
6. Shivers, B. D., Harlan, R. E., Pfaff, D. W., and Schachter, B. S. (1986). *J. Histochem. Cytochem.*, **34**, 39.
7. Mullink, H., Walboomers, J. M. M., Tadema, T. M., Jansen, D. J., and Meijer, C. J. L. M. (1989). *J. Histochem. Cytochem.*, **37**, 603.
8. Van der Loos, C., Volkers, H., Rook, R., Van den Berg, F., and Houthoff, H. J. (1989). *Histochem. J.*, **21**, 279.
9. Mullink, H., Jiwa, N. M., Walboomers, J. M. M., Horstman, A., Vos, W., and Meijer, C. J. L. M. (1991). *Am. J. Dermatopathol.*, **13**, 530.
10. Mullink, H., Boorsma, D. M., Henzen-Logmans, S. C., and Meijer, C. J. L. M. (1987). In *Application of monoclonal antibodies in tumor pathology* (ed. D. J. Ruiter, G. J. Fleuren, and S. O. Warnaar), pp. 37–47. Martinus Nijhof Publ., Amsterdam.
11. Cromme, F. V., Meijer, C. J. L. M., Snijders, P. J. F., Uyterlinde, A., Kenemans, P., Helmerhorst, Th., Stern, P. L., Van den Brule, A. J. C., and Walboomers, J. M. M. (1993). *Br. J. Cancer*, **67**, 1372.
12. Jiwa, N. M., Kanavaros, P., de Bruin, P. C., Van der Valk, P., Horstman, A., Vos, W., Mullink, H., Walboomers, J. M. M., and Meijer, C. J. L. M. (1993). *J. Pathol.*, **170**, 129.
13. Wolber, R. A. and Lloyd, R. V. (1988). *Hum. Pathol.*, **19**, 736.
14. Rentrop, M., Knapp, B., Winter, H., and Schweizer, J. (1986). *Histochem. J.*, **18**, 271.
15. McLean, I. W. and Nakane, P. K. (1974). *J. Histochem. Cytochem.*, **22**, 1077.
16. Hsu, S. M., Raine, L., and Fanger, H. (1981). *J. Histochem. Cytochem.*, **29**, 577.
17. Hsu, S. M. and Soban, E. (1982). *J. Histochem. Cytochem.*, **30**, 1079.
18. Merchentaler, I., Stancovics, J., and Gallyas, F. (1989). *J. Histochem. Cytochem.*, **37**, 1563.
19. Cordell, J. L., Falini, B., Erver, W. N., Ghosh, A. K., Abdulaziz, Z., MacDonald, S., Pulford, K. A. F., Stein, H., and Mason, D. Y. (1984). *J. Histochem. Cytochem.*, **32**, 219.
20. Bondi, A., Chieregatti, G., Eusebi, V., Fulcheri, E., and Bussolati, G. (1982). *Histochemistry*, **76**, 153.
21. Brigati, D. J., Myerson, D., Leary, J. J., Spalholz, B., Travis, S. Z., Fong, C. K. Y., Hsiung, G. D., and Ward, D. C. (1983). *Virology*, **126**, 32.
22. Hayashi, S., Gillam, I. C., Delaney, A. D., and Tener, G. M. (1978). *J. Histochem. Cytochem.*, **26**, 667.
23. Walboomers, J. M. M., Melchers, W. J. G., Mullink, H., Meijer, C. J. L. M., Struyk, A., Quint, W. G. J., Van der Noordaa, J., and ter Schegget, J. (1988). *Am. J. Pathol.*, **131**, 587.
24. Burns, J., Graham, A. K., Frank, C., Fleming, K. A., Evans, M. F., and McGee, J. O. D. (1987). *J. Clin. Pathol.*, **40**, 858.
25. McGadey, J. (1970). *Histochemie*, **23**, 180.
26. Mullink, H., Vos, W., Jiwa, M. N., Horstman, A., Van der Valk, P., Walboomers, J. M. M., and Meijer, C. J. L. M. (1992). *J. Histochem. Cytochem.*, **40**, 495.

27. Brahic, M. and Ozden, S. (1992). In *in situ hybridization: a practical approach* (ed. D. G. Wilkinson), pp. 85–104. IRL Press, Oxford.
28. Pringle, J. H., Ruprai, A. K., Primrose, L., Keyte, J., Potter, L., Close, P., and Lauder, I. (1990). *J. Pathol.*, **162**, 197.
29. Gillitzer, R., Berger, R., and Moll, H. (1990). *J. Histochem. Cytochem.*, **38**, 307.

6

Non-isotopic detection of nucleic acids on membranes

J. R. HUGHES, M. F. EVANS, and E. R. LEVY

1. Introduction

Membrane-based methods of DNA and RNA analysis have traditionally relied on the use of radiolabelled probes. Whilst they offer a robust and sensitive system there are associated problems involving the instability of the isotope, the requirement for designated laboratory areas and equipment for the handling of the isotope, and the specialist removal of waste. These problems have led to the development of a number of non-isotopic techniques for the analysis of nucleic acids that are sensitive enough to detect unique copy sequences within genomic digests.

Non-isotopic alternatives are classified into two categories (indirect and direct) depending on the method of labelling the probe. The majority of systems available for filter assays involve indirect detection of a labelled probe in several stages (*Figure 1*):

(a) Label the sequence.

(b) Hybridize the sequence to membrane-bound nucleic acid.

(c) Detect nucleic acid: probe hybrids *via* the specific binding of a non-radioactive reporter system.

The reporter system involves the interaction between the labelled probe and either an antibody or a specific binding protein coupled to an enzyme such as alkaline phosphatase (AP), glucose-6-phosphate dehydrogenase or horseradish peroxidase (HRP). The enzyme reacts with a specific substrate to produce either a coloured, luminescent, or fluorescent product. Colourimetric techniques involve the formation of an insoluble coloured precipitate on the membrane (Section 6.3.1). Chemiluminescent systems involve the chemical production of light. The procedures are similar to those used in colour detection with an end-point of light emission at a wavelength that is detected by X-ray film, giving a result visually similar to that of radiolabelled probes (Section 6.3.2). Fluorescent techniques are less commonly used for

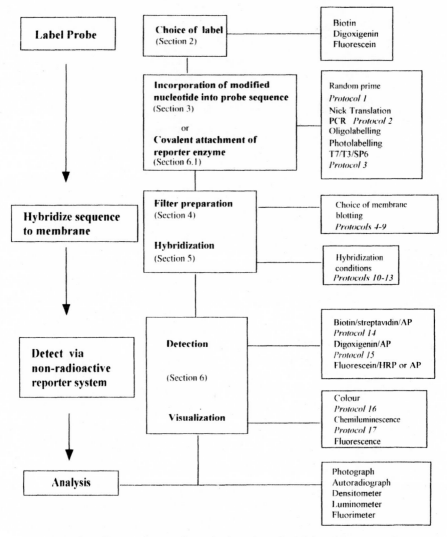

Figure 1. Flow diagram for non-isotopic detection of nucleic acids on membranes.

the analysis of nucleic acids on membrane supports since they rely on specialist equipment for the detection and analysis of the signal.

Direct methods involve covalent attachment of the signalling enzyme to the probe. This removes the need to use an antibody or binding protein which in turn eliminates non-specific binding and reduces background signal. This technique also reduces the detection time. A variety of non-isotopic techniques for nucleic acid analysis are reviewed in detail in ref. 1.

In this chapter we describe the indirect analysis of non-radioactive probes, including some protocols for probe labelling, and review the handling of nucleic acids on to membrane filters and their subsequent hybridization. *Figure 1* gives a flow diagram of the detection methodologies available which culminate in a colorimetric or chemiluminescent end-point. This chapter also covers trouble-shooting in terms of signal–noise optimization considerations. Finally, at the end of the chapter we list a variety of the commercial kits available.

2. Choice of label

2.1 Biotin

One of the first reported non-radioactive methods for nucleic acid analysis used the vitamin biotin as part of a reporter group (2, 3). The probe sequence is modified with biotin and hybridized to the nucleic acid under study (see Section 6.1).

2.2. Digoxigenin

The digoxigenin (DIG) system was developed to avoid the disadvantages arising from the use of a ubiquitous vitamin such as biotin (4). The probe is modified with an artificial cardenolide-hapten digoxigenin, occurring specifically in the *Digitalis* plant, thus avoiding non-specific binding to tissue specimens (see Section 6.1).

2.3 Fluorescein

Probes are labelled with fluorescein-11-dUTP and detected with an enzyme-linked anti-fluorescein antibody (Section 6.3.2).

3. Probe labelling

A major advantage of non-isotopic labelling is the stability and ease of handling of the probes. As there is no short half-life as seen with ^{32}P, probes can be labelled in bulk and stored at $-20°C$. This is particularly appropriate for PCR-generated probes. Additionally, with the safety requirements usually observed for radioactive technology removed, it is possible to label large numbers of probes at one session. Various methods of probe labelling are reviewed in Chapter 1 of this volume.

3.1 DNA

The most common method for filter-based assays involves incorporation of a suitably modified nucleotide into the probe. Other methods such as direct labelling to incorporate the reporter enzyme are described in Chapter 1.

3.1.1 Random prime labelling

The most common methods for labelling nucleic acids for filter hybridizations are nick-translation (5) and random-priming (6), for the incorporation of digoxigenin, biotin, or fluorescein-labelled deoxyuridine. Commercial kits are available for all these labels (e.g. Boehringer–Mannheim, Gibco BRL, Amersham International Ltd). *Protocol 1* describes random-prime labelling.

Protocol 1. Random-prime labelling of DNA with digoxigenin–dUTP [a]

Equipment and reagents

- 10 × hexanucleotide mixture (Boehringer–Mannheim): 62.5 A_{260} units/ml random hexanucleotides, 500 mM Tris–HCl, pH 7.2, 100 mM $MgCl_2$, 1 mM dithioerythritol (DTE), 2 mg/ml BSA
- 10 × digoxigenin (DIG) DNA labelling mixture (Boehringer–Mannheim): 1 mM dATP,
- 1 mM dCTP, 1 mM dGTP, 0.65 mM dTTP, 0.35 mM DIG-11-dUTP, pH 7
- Klenow enzyme labelling grade (Boehringer–Mannheim) 2 U/µl
- Select-D columns (5'-3' inc. CP Laboratories)
- distilled water
- purified insert DNA

Method

1. Mix the following in a clean microcentrifuge tube:
   - purified insert DNA   100–200 ng
   - sterile distilled H_2O to a final volume of   10 µl

   heat to 95°C for 5 min then cool on ice for 5 min.

2. Add the remaining reagents:
   - hexanucleotide mixture   2 µl
   - labelling mixture   2 µl
   - Klenow   1 µl

   mix gently and incubate for 1–2 h at 37°C.

3. Remove unincorporated nucleotides by spin dialysis using Select-D columns according to the manufacturer's instructions.

4. Check incorporation (Section 7.4) or proceed to hybridization (Section 5).

5. Probes may be stored at −20°C until required.

[a] DNA may be labelled with biotin using a Megaprime™ labelling kit (Amersham International Ltd) and 400 pmol Biotin-16-dUTP (Sigma, Boehringer–Mannheim). Remove unincorporated nucleotides by spin dialysis with Select-B (5'-3' inc.) columns. (These columns have been equilibrated with SDS to prevent the binding of the biotinylated probe to the column material.)

3.1.2 Polymerase chain reaction (PCR)

It is possible to generate large amounts of labelled insert from probes in pUC and other vectors using M13 sequencing primers and the polymerase chain reaction (7–9) by replacing the unlabelled nucleoside triphosphate mix with one containing DIG–dUTP or biotin–dUTP (*Protocol 2*). This technique is most useful for small inserts as full-length labelled probes are produced. RNA probes may be labelled in a similar way with a reverse transcription step prior to amplification.

Protocol 2. Labelling of DNA in pUC [a] vectors by the PCR

Reagents

- M13 forward and reverse primers (working stock of 40 pmol): forward 5'-GTAAAACGACGGCCAGT-3' reverse 5'-CAGGAAACAGCTATGAC-3'
- ultrapure dNTP set of 2'-deoxynucleoside 5'-triphosphates (Pharmacia)
- DIG-11-dUTP (Boehringer–Mannheim)
- biotin-16-dUTP (Boehringer–Mannheim)
- *Taq* DNA polymerase (Boehringer–Mannheim)

- 2 × PCR buffer: 20 mM Tris–HCl, pH 8.5, 100 mM KCl, 3 mM $MgCl_2$ [b], 0.02% gelatin
- distilled water
- mineral oil
- probe DNA in pUC vector
- All reagents in the PCR buffer should be autoclaved or filter-sterilized.

Method

1. Set up 100 μl reactions in sterile microcentrifuge tubes [c]:

   - each primer   20 pmol
   - 2 × PCR buffer   50 μl
   - dATP, dCTP, and dGTP   0.2 mM each
   - dTTP   0.13 mM
   - DIG–dUTP or biotin–dUTP   0.07 mM [d]
   - plasmid DNA   0.1–0.15 ng
   - *Taq* polymerase   2.5 units
   - sterile distilled water   to 100 μl

2. Mix well, microcentrifuge for 5 sec, and overlay with 100 μl mineral oil.

3. Carry out the following cycle sequence [e]:

   - initial denaturation at 94°C for 5 min followed by 30 cycles of:
   - denaturation at 94°C for 1.5 min
   - anneal at 50°C for 1.5 min
   - extension at 72°C for 2.5 min
   - final elongation step at 72°C for 10 min.

4. Following amplification remove the mineral oil and purify the product through a Select-B or Select-D column (*Protocol 1*).

Protocol 2. *Continued*

5. Run 5 μl on an agarose gel to confirm that the product is the correct size and uncontaminated. This will also allow the concentration to be assessed (see also Section 7.4). Note that the addition of the hapten will cause the insert to migrate slower than an unlabelled insert and it will, therefore, appear slightly larger.

6. Probes may be stored at −20°C until required.

> [a] Primers for other vectors are also available.
> [b] To achieve optimum yields of the probe the MgCl₂ concentration should be assayed for each probe.
> [c] A control reaction containing no DNA should be set up alongside to check for contamination.
> [d] The ratio of DIG–dUTP and biotin–dUTP to dTTP may be varied to optimize incorporation (10).
> [e] For a detailed protocol of PCR generation of probes see ref. 7.

3.1.3 Oligonucleotide labelling

Oligonucleotides may be labelled with terminal deoxynucleotidyl transferase adding either a single hapten to the 3′ end or by tailing the probe with a mixture of nucleotides. These techniques provide a very efficient method of labelling small probes (Chapter 1 of this volume). Kits are also available for direct labelling with alkaline phosphatase (Section 8).

3.1.4 Photolabelling

Probes may be labelled with Photodigoxigenin (Boehringer–Mannheim) or Photobiotin™ (Gibco BRL) which results in intact molecules. This is very useful for the generation of labelled size markers since the size of the labelled fragment remains unaltered. The technique results in a low-labelling density which allows the labelled DNA to migrate at the same velocity as unlabelled DNA (9). However, since the labelling density is so low, the sensitivity is correspondingly reduced and it is, therefore, not an appropriate technique for the labelling of oligonucleotides or single-copy probes.

3.2 RNA

3.2.1 *In vitro* transcription

This procedure utilizes plasmid vectors containing *E. coli* bacteriophage RNA polymerase promoters such as SP6, T3, or T7. Many such vectors are commercially available (e.g. the pGEM series, Promega), and are constructed such that a multiple cloning site (MCS) is sandwiched between promoter sites. The DNA of interest is engineered into the MCS using appropriate restriction enzymes. This construct is subsequently linearized depending on whether 'sense' or 'antisense' RNA is required. Following complete digestion the template is purified by standard phenol/chloroform extraction procedure and ethanol precipitation. *In vitro* transcription is then carried out as described in *Protocol 3*.

Protocol 3. RNA labelling by *in vitro* transcription

Reagents

Always use RNase free, DEPC-treated reagents [a]

- linearized template DNA (see Section 3.2.1)
- 10 × transcription buffer (10 × TB): 20 mM spermidine (Sigma), 60 mM MgCl$_2$, 400 mM Tris–HCl, pH 7.5, 100 mM NaCl, 100 mM DTT (Sigma)
- placental ribonuclease inhibitor, RNasin (Pharmacia): 40 U/µl
- RNase-free BSA (Sigma): 2 mg/ml (optional)
- SP6, T3, or T7 polymerase: 10–20 U/µl
- 0.25 M EDTA, pH 8.0

- 4 M LiCl
- absolute ethanol, 70% ethanol
- *either* 10 × DIG-rNTPs labelling mix: 10 mM each rATP, rGTP, rCTP, 6.5 mM rUTP, 3.5 mM DIG–UTP (Boehringer–Mannheim), 10 mM Tris–HCl, pH 7.5
- *or* 10 × biotin-rNTP labelling mix: 10 mM each rATP, rGTP, rCTP, 6.5 mM rUTP, 3.5 mM biotin–UTP (Clontech, Sigma), 10 mM Tris–HCl, pH 7.5

Method

1. Mix the components in the following order [b] in a microcentrifuge tube:
   - 10 × TB   2 µl
   - water   11 µl
   - RNasin   1 µl
   - linearized template DNA   1 µl (1 µg)
   - 10 × DIG-rNTPs or 10 × biotin-rNTPs   2 µl
   - RNase free BSA   1 µl (optional)
   - RNA polymerase   2 µl (40 U)

2. Incubate at either 37°C for 2 h (T3, T7 polymerase), or 40°C for 2 h (SP6 polymerase).

3. Stop the reaction by the addition of 2 µl 0.25 M EDTA, pH 8.0.

4. Add 2 µl 4 M LiCl and 3 volumes of ethanol pre-chilled to −20°C.

5. Precipitate overnight at −20°C or −70°C for > 1 h.

6. Centrifuge at 1300 *g* for 15 min (preferably at 4°C).

7. Pour off the supernatant and wash the pellet with 100 µl of 70% ethanol chilled to −20°C.

8. Centrifuge at 1300 *g* for 5 min and carefully remove the 70% ethanol.

9. Dry the pellet, dissolve in 100 µl DEPC-treated water and add 1 µl RNasin.

10. Estimate the labelling efficiency [c] by Dot blot detection (Section 7.4), or detect the label following glyoxal denaturing electrophoresis and Northern transfer (*Protocol 7*). [d]

Protocol 3. *Continued*

11. Aliquot and store at $-20\,^{\circ}\mathrm{C}$ or $-70\,^{\circ}\mathrm{C}$.

---

[a] Add DEPC to a concentration of 0.1% to stock solutions of $MgCl_2$, NaCl, LiCl, EDTA, and to water. Leave overnight and then autoclave. Prepare Tris buffers with autoclaved DEPC-treated water and then autoclave.

[b] Spermidine concentrations greater than 4 mM may precipitate the template DNA, so dilute the TB first.

[c] Incorporation of biotin–UTP into RNA by SP6 polymerase is low (11).

[d] Use 1 μl of probe for Dot blot assessment, 2.5 μl for gel electrophoretic analysis

4. Filter preparation

4.1 Membrane selection

The selection of a suitable membrane depends on the system being used for detection. For a colorimetric assay, nitrocellulose membranes are best, although suitable nylon membranes are commercially available (Boehringer–Mannheim, Oncor). For chemiluminescence reactions, nylon membranes with a low positive charge are most suitable for digoxigenin-labelled probes. The use of membranes with a higher positive charge may result in a higher non-specific background. Biotinylated probes also cause non-specific background, but membranes such as Immobilon-S (Millipore) are available for these techniques. Nitrocellulose membranes are unsuitable for chemiluminescence unless used in conjunction with a specific blocking agent Nitro-Block (Tropix) (Section 7.6).

4.2 Southern blots

The preparation of DNA samples for standard analysis and pulsed-field studies have been described in detail in refs 12, 13.

4.2.1. Gel preparation

Protocol 4 summarizes one method of gel preparation for Southern blotting. For more details see refs 12–14.

Protocol 4. Gel preparation

Reagents

- 10 × TBE; 0.89 M Tris–borate, 0.02 M EDTA.Na, pH 8.3 [a]
- agarose (Sigma)
- Ficoll (Sigma)
- bromophenol blue (Sigma)
- EDTA.Na (0.5 M)
- Sodium dodecyl sulfate (SDS) 10% (w/v) in dH_2O, pH 7.2
- Ficoll loading buffer: 7.5 g Ficoll, 0.05 g Bromophenol Blue, 5 ml EDTA, 1.25 ml SDS, 25 ml 10 × TBE, heat and stir, make up to 50 ml with dH_2O, pH 8.5
- DNA samples

Method

1. Prepare a 0.8–1% agarose gel of required size, i.e. for 100 ml gel:
   - agarose   0.8 g
   - 1 × TBE   100 ml

2. Heat in a microwave until the solution is completely clear. Make up to 100 ml with dH_2O.

3. Cool to 50–60°C and pour into a casting mould with the appropriate comb. Leave to set.

4. Fill the electrophoresis tank with running buffer (1 × TBE).

5. Prepare the DNA samples as follows (for 20 μl samples):
   - loading buffer   4 μl
   - digested DNA x μl
   - dH_2O to 20 μl

   microcentrifuge the samples for 5 sec.

6. Prepare the appropriate markers (Section 4.2.2) as in Step 5.

7. Transfer the casting tray and gel when set into the tank, remove the comb, and load the samples. Run until the blue dye has reached the required distance in the gel. The voltage will depend on the size of the gel and run time.

---

[a] An alternative buffer is Tris–Acetate–EDTA (TAE). A 10 × stock solution consists of 0.4 M Tris base, 0.2 M sodium acetate (trihydrate), 2 mM EDTA, pH to 8.2 with glacial acetic acid.

4.2.2 Markers

A range of biotinylated size markers ready to load on to the gel are available from Gibco BRL, and DIG-labelled or biotinylated markers from Boehringer–Mannheim. It is also possible to make a stock of digoxigenin or biotinylated lambda phage DNA by nick-translation or random-prime labelling. The resulting labelled marker should be diluted to 0.5–1 ng/μl and stored at −20°C. Prior to hybridization, 1 μl is added to the probe mixture which should hybridize to unlabelled marker run on the gel. This technique is particularly appropriate for pulsed-field gels where lambda concatamers are used as size markers.

4.2.3 Transfer

DNA is transferred to the membrane of choice by standard procedures (15) (*Protocol 5*). The membranes commonly used in our laboratory for chemiluminescence detection with DIG-labelled probes are Hybond-N (Amersham International Ltd) or Nytran (Scleicher and Schuell). However, there is marked variation between batches resulting in variable signal to background ratio. For colorimetric detection BA85 nitrocellulose (Scleicher and Schuell) and Biotrace NT (Gelman) have been used.

Protocol 5. Transfer of DNA to membrane

Equipment and reagents

- ethidium bromide: 10 mg/ml stock solution
- denaturing solution: 1 M NaOH
- Neutralizing buffer: 1 M Tris–HCl, 3 M NaCl, pH 7.5
- 10 × SSC, pH 7.0: 1.5 M sodium chloride, 0.15 M sodium citrate
- membrane

Method

1. Visualize the gel with ethidium bromide and UV-irradiate the gel.
2. Incubate for 1 h at room temperature with gentle shaking in denaturing solution.
3. Incubate for 1 h at room temperature in neutralization buffer (pH 7.5).
4. Transfer to Hybond-N or Nytran in 10 × SSC by capillary action [a] (15). It has been reported that transfer in 1 M ammonium acetate (NH_4Ac) results in a stronger signal for chemiluminescence (16).
5. Fix the DNA to the membrane according to manufacturer's instructions.
6. Store dry.

[a] Apparatus is commercially available for vacuum blotting, positive pressure blotting, and electroblotting.

4.3 DNA Dot blots

In some instances it is more convenient and useful to blot a liquid DNA sample directly on to the membrane without prior size fractionation through a gel, for example for quantification of a piece of DNA relative to standards (*Protocol 6*).

Protocol 6. DNA Dot blots

Equipment and reagents

- target DNA
- membrane

Method

1. Prepare a suitable range of dilutions of target DNA.
2. Denature by heating to 95–100°C for 10 min then snap cool on ice. [a]

3. Pipette 1–2 μl of the denatured dilution on to the membrane which has been prepared according to manufacturer's instructions and with appropriate orientation marks.

4. Transilluminate or bake the filter to fix the DNA.

[a] If a vacuum slot blotter is used samples should be denatured in NaOH.

4.4 Libraries

Filters for library screening are prepared by the same techniques as for radiolabelled probes (12). Problems may occur after chemiluminescence detection if background spotting is observed. It is essential, therefore, to make duplicate filters to be hybridized with the same probe to avoid false-positives (Section 7.3).

4.5 Northern blots

RNA samples may be prepared as described in ref. 17. Native RNAs have significant secondary structure which interferes both with their mobility through gels and their binding to membrane filters. The most widely used denaturing methods are those based on glyoxal and DMSO (*Protocol 7*) (18), and those using formaldehyde and formamide (*Protocol 8*) (19). RNA may also be denatured using the highly toxic methyl mercuric hydroxide (20). This method will not be described here.

4.5.1 RNA denaturation using glyoxal/dimethyl sulfoxide (DMSO)

This technique may be preferable to formaldehyde/formamide methods as the RNA bands subsequently detected following hybridization have a sharper appearance (12). This is an important consideration for the optimization of non-isotopic detection sensitivity. A suitable procedure is given in *Protocol 7*.

Protocol 7. RNA denaturation using glyoxal and dimethyl sulfoxide (DMSO)

Equipment and reagents

All reagents and apparatus should be RNase-free or DEPC-treated (see below).

- 6 M glyoxal (Sigma): deionize the 40% solution by passage through a mixed bed resin (e.g. Bio-Rad AG 501-X8, or MO32, Sigma) until the pH > 5. Store at −20°C in tightly capped bottles filled to capacity then discard after use.
- dimethylsulfoxide (DMSO): use high grade DMSO directly without pre-treatment. Aliquot new stocks and store at −20°C. Use each aliquot once.
- 10 × electrophoresis buffer: 100 mM sodium phosphate, pH 7.0 (3.9 ml of 1 M

Protocol 7. *Continued*

sodium dihydrogen phosphate, 6.1 ml of 1 M disodium hydrogen phosphate, 90 ml of DEPC-treated water)

- denaturing mixture: 3 ml deionized 6 M glyoxal, 7.5 ml DMSO, 1.5 ml 10 × electrophoresis buffer. Aliquot into small tubes and use once.

- glyoxal/DMSO gel loading buffer: 50% glycerol, 1 × electrophoresis buffer, 0.25% bromophenol blue, 0.25% xylene cyanol FF
- agarose – electrophoresis grade (Sigma)

A. *DEPC treatment*

RNA is a highly labile molecule requiring care in its preparation, use, and storage. By contrast, ribonucleases are stable and widespread in the environment being present in skin secretions for example. All reagents purchased should be RNase free, and any prepared solutions treated with diethylpyrocarbonate (DEPC) which largely eliminates ribonuclease activity.

1. Add the DEPC to the sodium phosphate solution and to quantities of water to a concentration of 0.1%. Mix well and allow to stand overnight in a fume cupboard (DEPC is toxic). Keep the tops of the bottles slightly loose to allow the CO_2 to escape. Next day autoclave the liquids to eliminate the DEPC.

2. Rinse the peristaltic pump (see below) with DEPC-treated water.

3. Prepare the electrophoresis tank by cleaning with detergent, rinse with water, dry with alcohol, and leave to stand for 10 minutes filled with 3% hydrogen peroxide. Finally rinse with DEPC-treated water.

B. *Gel preparation*

1. Prepare gel of appropriate length and agarose concentration[a] by adding agarose to 1 × electrophoresis buffer and heating in microwave or waterbath. Pour the gel when cooled to < 50°C. Solid sodium acetate may be added to a final concentration of 10 mM when the gel is at 70°C to further ensure RNase inactivation.

2. Add 3 µl (up to 30 µg) of total cellular or poly(A)[+] RNA[b] dissolved in DEPC-treated water to 12 µl denaturing mix.

3. Incubate at 50°C for 60 min.

4. Place the samples on ice for 2 min. Pulse-spin, add 2.5 µl gel loading buffer and immediately load.

5. Run the gel at 3–4 V/cm using 10 mM sodium phosphate as running buffer. Re-circulate the buffer from cathode to anode using a peristaltic pump.[c]

6. Load 1–2 µg of similarly denatured RNA markers and/or 10–100 ng labelled RNA markers.[d]

7. Run the gel until the bromophenol blue has migrated 80% of the gel length (to within 1–2 cm of the end of the gel)

8. Cut the lanes containing any unlabelled markers from the gel and stain them by soaking in ethidium bromide[e] (0.5 μg/ml in 0.1 M ammonium acetate) for 30 min. Photograph the gel aligned with a clear ruler and illuminated by UV light[f]. De-stain if necessary by immersion in water or 1 mM Mg_2SO_4.

9. Wash the remaining gel for 5 min using two changes of 10 mM sodium phosphate then proceed with Northern transfer (Section 4.5.3).

[a] Use a 1.4% gel 10–15 cm long to fractionate RNA molecules up to 1 kb in length, and for longer molecules use a 1% gel.

[b] For detection of known abundant mRNAs (0.1% or more of the mRNA population) use > 10 μg of total cellular mRNA. 2–5 μg of poly (A)[+] RNA is suggested for detection of rarer RNAs.

[c] It is essential that RNA samples are completely denatured before electrophoresis and remain so subsequently. Recirculating the buffer maintains the pH < 8.0, above which glyoxal dissociates from RNA. Should the dye turn green/yellow at any stage stop the electrophoresis, change the buffer and continue when the dye is blue again. It has also been reported (17) that 1 × MoPS (see *Protocol 8*) may be used as running buffer without the need for recirculation.

[d] Digoxigenin-labelled RNA markers are available from Boehringer–Mannheim. Biotinylated DNA markers are available from Gibco/BRL Life Technologies and Oncor. Glyoxylated DNA migrates similarly to glyoxylated RNA and so DNA markers may be used (21).

[e] Post-electrophoresis staining is necessary as the reaction between the RNA and ethidium bromide interferes with fractionation. **Note**: ethidium bromide is a poweful mutagen requiring careful handling and disposal.

[f] Plot, on semi-log paper, the size of the RNA markers (log scale) against distance migrated. The resulting line can be used to estimate the sizes of the RNA molecules detected following transfer and hybridization.

4.5.2 RNA denaturation using formaldehyde and formamide

Gels containing formaldehyde are easier to run than those of glyoxal/DMSO-treated RNA and for this reason are more popular. Both formaldehyde and formamide are toxic. These gels should be prepared and run in a fume hood. *Protocol 8* describes the methodology.

Protocol 8. RNA denaturation using formaldehyde and formamide

Reagents

Note: Formaldehyde and formamide are toxic. These gels should be prepared and run in a fume hood

- formamide: high-quality formamide can generally be used without pre-treatment; deionization is required if a yellow colour is observed (*Protocol 7*)
- formaldehyde: usually obtained as a 37% solution (12.3 M) in water. Keep in tightly closed bottles to prevent oxidation. Discard if the pH of the concentrated solution is <4.0.
- 5 × formaldehyde gel running buffer: 0.1 M MoPS, pH 7.0, 40 mM sodium acetate, 5 mM EDTA. Dissolve 20.6 g of 3-(N-morpholino) propanesulfonic acid (Mops), in 800 ml

DEPC-treated (*Protocol 7*) sodium acetate. Adjust the pH to 7.0 with 2 M NaOH and add 10 ml of DEPC-treated 0.5 M EDTA, pH 8.0. Add DEPC-treated water to 1 litre. Filter-sterilize the solution through a 0.2 μm membrane and store at room temperature sealed from light. Straw-coloured buffer is useable, a dark buffer is not.
- 20 × SSC: 3 M sodium chloride, 0.3 M sodium citrate, pH 7.0
- formaldehyde loading buffer: 50% glycerol, 1 m M EDTA, pH 8.0, 0.25% bromophenol blue, 0.25% xylene cyanol FF

Protocol 8. *Continued*

Method

1. Prepare an appropriate gel (see *Protocol 7*). When the gel is at approx. 60°C, add 5 × formaldehyde gel running buffer to a final concentration of 1 ×. Add formaldehyde to a final concentration of 0.66 M[a] (for example for 100 ml: 5.4 ml formaldehyde solution, 74.6 ml 1.4% agarose/DEPC-water, 20 ml 5 × buffer — final agarose concentration is 1.0%).

2. Prepare the samples by mixing in a sterile microcentrifuge tube:
   - RNA (up to 30 µg)   4.5 µl
   - 5 × buffer   2.0 µl
   - 37% formaldehyde   3.5 µl
   - formamide   10 µl

3. Denature the samples by heating for 15 min at 65°C then chill on ice for 5 min. Microcentrifuge briefly.

4. Add 2 µl sterile DEPC-treated formaldehyde gel-loading buffer and briefly microcentrifuge.

5. Pre-run the gel in 1 × gel running buffer for 5 min at 5 V/cm. Load the samples and the markers (*Protocol 7*) into the wells.

6. Continue electrophoresis at 1–3 V/cm.

7. On termination of electrophoresis remove the formaldehyde from the gel by rinsing in several changes of 1 × MoPS or DEPC-treated water for 15 min. Then proceed with Northern transfer (Section 4.5.3).

---

[a] A final formaldehyde concentration of 2.2 M is recommended in older protocols, but more recent reports suggest a concentration of 0.66 M is just as effective (17) and is safer.

4.5.3 Northern transfer

The transfer of RNA from gels to membrane is essentially the same as for Southern blotting (Section 4.2.3) (15) using 20 × SSC and capillary transfer on to nitrocellulose membranes for 12–16 h, but with no requirement for NaOH denaturation and neutralization (18). The transfer apparatus should be cleaned in a similar way to the gel tanks (*Protocol 7*).

4.6 RNA Dot blots

This technique, which is similar to that of DNA Dot blots, is rapid, sensitive, and can be semi-quantitative.

Protocol 9. Preparation of RNA Dot blots

Equipment and reagents

All reagents should be RNase free or DEPC-treated

- 6 M glyoxal
- formaldehyde, 37% solution
- formamide
- 20 × SSC
- membrane (see *Protocols 7* and *8* for preparation)

Method

1. Cut a piece of nitrocellulose or nylon filter to the desired size. Wet with water then soak in 20 × SSC.

2. Denature the RNA samples with glyoxal *or* formaldehyde as follows:

    (a) Glyoxal denaturation — combine the following:

    - RNA solution (up to 20 µg)   3 µl
    - 6 M glyoxal   2 µl
    - 20 × SSC   5 µl

    heat at 50°C for 1 h

    (b) Formaldehyde/formamide denaturation — combine the following:

    - RNA solution (up to 20 µg)   2 µl
    - formamide   5 µl
    - formaldehyde   2 µl
    - 20 × SSC   1 µl

    heat at 65°C for 1 h

3. Chill the samples on ice for 5 min. Microcentrifuge briefly.

4. Place the nitrocellulose or nylon membrane on to a filter paper soaked in 20 × SSC supported on Cling-film.

5. When the membrane appears moist and drip-free, spot the sample out in successive 2–5 µl drops so as to limit dot surface area.[a]

6. Transfer the membrane to a dry filter paper, air-dry briefly, rinse in 20 × SSC briefly, then bake at 80°C or treat with UV irradiation (nylon only) using a Stratolinker (Stratagene) or callibrated UV illuminator. The filters are then ready for hybridization or storage at 4°C. Store nitrocellulose under vacuum or in 20 × SSC at 4°C.

[a] Alternatively, use a commercially available Dot/Slot apparatus according to the instructions supplied. A Slot gives a higher signal intensity than a Dot and is recommended where densitometric quantification is desired.

---

5. Hybridization of filters

5.1 DNA

5.1.1 Hybridization solutions

Reaction conditions are reported to be similar to those of isotopically-labelled probes although the melting temperature (T_m) of biotinylated probes is lower (3). Therefore, there is little need to modify existing protocols. Hybridization has been performed in a variety of standard phosphate and formamide mixtures (12) with the addition of extra single-stranded sonicated salmon sperm DNA to decrease non-specific background. A list of the reagents for different hybridization solutions applicable to Southern blots is shown below.

i. Reagents for hybridization solutions
Formamide-based pre-hybridization and hybridization solution
- formamide AR (Fluka)
- dextran sulfate (Pharmacia)
- sodium dodecyl sulfate (SDS), pH 7.2 (Sigma, BDH)
- 50 × Denhardt's solution: 1% (w/v) Ficoll; 1% (w/v) polyvinylpyrrolidone; 1% (w/v) BSA, Fraction V (Sigma)
- 20 × SSC: 3 M sodium chloride, 0.3 M sodium tricitrate, pH 7.0
- salmon sperm DNA, sonicated to approx. 500 bp: stock 10 mg/ml in dH_2O

Use these reagents to prepare the following pre-hybridization/hybridization solution: 50% formamide, 6 × SSC, 5% dextran sulfate, 1% SDS, 5 × Denhardt's solutions (see **Note**) with 100 μg/ml sonicated single-stranded salmon sperm DNA added just prior to use.

Phosphate-based hybridization solution
- dextran sulfate (Pharmacia)
- sodium dodecyl sulfate (SDS)
- 50 × Denhardt's solution: see above
- 20 × SSPE: 3.6 M NaCl, 200 mM NaH_2PO_4, 20 mM EDTA, pH 7.4.
- salmon sperm DNA as above.

Use these reagents to prepare the following pre-hybridization/hybridization solution:

5 × Denhardt's solution,[a] 5 × SSPE, 0.3% SDS, 10% dextran sulfate, 100 μg/ml sonicated single-stranded salmon-sperm DNA.

Church and Gilbert (12) hybridization mixture
This has also been used successfully with non-isotopic probes with the addition of 100 μg/ml sonicated single-stranded salmon sperm.

Note: a 1–2% solution of blocking reagent (Boehringer–Mannheim) may be preferable for colourimetric studies (p. 176).

5.1.2 Hybridization in the presence of formamide

Protocol 10 has been used to detect single copy sequences in genomic DNA digests with DIG-labelled probes, detected using a chemiluminescent reaction. It is also applicable to biotinylated probes. The phosphate-based hybridization solution is better for cross-species studies as it is of a lower stringency. Where radioactive probes have not been used previously *Protocol 10* is a good place to start; otherwise it is sensible to modify existing protocols.

Protocol 10. Hybridization in the presence of formamide

Equipment and reagents
- pre-hybridization solution (see Section 5.1.1.*i*.)
- salmon sperm
- hybridization bags
- competitor DNA (see Section 5.1.3)
- marker DNA (see Section 4.2.2)

A. *Pre-hybridization*

1. Warm the pre-hybridization mixture to 50°C to ensure all components dissolve.
2. Denature the sonicated salmon sperm DNA by heating to 95°C for 5 min then chill on ice. Add the freshly denatured salmon sperm to the pre-hybridization solution, mix and place in a hybridization bag with the filter (or two filters back to back) using approximately 20 ml/100 cm^2 of filter.
3. Incubate at 42°C for 1–2 h (65°C for the phosphate mix).

B. *Hybridization*

1. Denature the probe DNA with competitor (Section 5.1.3) and marker (Section 4.2.2) if required for 10 min at 95°C and chill on ice. Use 50 ng in 5 ml of hybridization solution for 100 cm^2 of membrane for standard filters and 80–100 ng for pulsed-field filters.
2. Remove the pre-hybridization solution from the bags. This may be re-used as the hybridization mixture.
3. Add the denatured probe to the pre-warmed buffer, mix well, and replace in bags.
4. Remove air bubbles, seal, and ensure the solution is well distributed over the filters.
5. Incubate overnight at 42°C (65°C for phosphate and Church and Gilbert mix).

5.1.3 Competitor

We have found that the addition of 100 µg/ml single-stranded salmon sperm DNA (sonicated to approx. 500 bp) to all hybridization buffers greatly reduces non-specific background, which is otherwise higher with non-isotopic detections compared to the same probe that has been radiolabelled. Repetitive sequences in probes, such as cosmids, may be competed out with the addition of 125–500 µg (depending on the amount of repetitive sequences) sonicated human DNA added to the probe mix prior to denaturation. Pre-incubation is not required.

5.1.4 Re-use of hybridization mixture

Following hybridization, the solution containing the probe can be retained for future use. This is particularly useful for screening large numbers of filters with the same probe. To re-use a formamide-based mixture denature the solution at a temperature between 60–70°C by standing in a waterbath. This must be carried out in a fume hood to avoid fumes from the hot formamide.

Hybridization mixtures without formamide require denaturation at a higher temperature. Place the solution in a boiling waterbath for 5–10 min and cool on ice just before use.

5.1.5 Post-hybridization washes

Filters should be washed under standard stringent conditions, for example *Protocol 11* can be used for unique sequence human probe to human genomic samples.

Protocol 11. Post-hybridization washes

Reagents

- 2 × SSC: 0.3 M sodium chloride, 30 mM sodium citrate
- 2 × SSC, 0.2% SDS
- 0.1 × SSC, 0.2% SDS

Method

1. Remove the filters from the bag into 2 × SSC at room temperature.
2. Wash in two changes of 2 × SSC, 0.2% SDS at room temperature for 5 min, each with shaking.
3. Wash in two changes of 0.1 × SSC, 0.2% SDS at 65°C for 15 min each (0.05 × SSC, 0.2% SDS has been found to work well for competition blots; S. Thomas, personal communication).

For untested probes it is possible to wash at low stringency initially, for example several rinses in 2 × SSC at room temperature followed by incubation in 2 × SSC at 65°C, then detection of the filter. If there is background on the film or excessive lane smear the filter can be re-washed to a higher stringency (by decreasing the concentration of the SSC) and re-detected (Section 7.2). Filters may be stored damp at 4°C for later detection. It is essential that filters are never allowed to dry out after hybridization.

5.2 RNA

5.2.1 Hybridization solutions

RNA hybridizations are generally performed in the presence of formamide which decreases the T_m (melting temperature) of the hybrid. This allows lower incubation temperatures to be used which are favoured because high temperatures may degrade the target RNA causing it to dissociate from the membrane filter. Exact temperature/stringency conditions can be calculated, but, in practise, some trial and error experimentation is also required. Hybridization conditions are thoroughly reviewed in refs 12, 22–24. A list of the solutions for RNA hybridization is given below, and *Protocol 12* describes the techniques involved.

i. RNA hybridization solutions

All reagents should be RNase-free or DEPC-treated (*Protocol 7*).

- 20 × SSPE: 3.6 M NaCl, 0.2 M sodium phosphate, 20 mM EDTA, pH 7.2
- *N*-laurylsarcosine, Na-salt
- sodium dodecyl sulfate (SDS)
- blocking reagent[a]
- formamide
- sonicated herring sperm DNA or carrier tRNA (Sigma)

Hybridization solution

Use these reagents to prepare the following hybridization solution:
5 × SSPE,[b] 0.1% (w/v) *N*-laurylsarcosine, Na-salt, 0.02% SDS, 2% blocking reagent, 50% formamide, 100 μg/ml denatured sonicated herring sperm DNA,[c] or 100–200 μg/ml carrier tRNA.

An alternative hybridization mixture recommended for use with DNA probes on 'nylon' Northern blots (25) is: 7% SDS, 50 mM sodium phosphate, pH 7.0, 50% formamide, 2% blocking reagent, 50 μg total yeast RNA, 5 × SSC, 0.1% laurylsarcosine.

---

[a] A purified dry milk fraction is available from Boehringer–Mannheim. Prepare as directed by the manufacturer to ensure the absence of RNase. In our experience use of this reagent is highly effective in reducing non-specific background staining especially when using a colorimetric detection protocol. 'Food' quality non-fat milk may contain significant RNase contamination.
[b] 5 × SSC may be used instead, but SSPE provides greater buffering capacity.
[c] Denature immediately prior to use by heating at 95°C for 10 min.

J. R. Hughes, M. F. Evans, and E. R. Levy

Hybridization buffer for oligonucleotide probes
5 × SSPE, 0.1%(w/v) *N*-laurylsarcosine, 0.02% (w/v) SDS, 1% blocking
reagent, 100–200 μg/ml tRNA or 100 μg/ml Poly(A), 5 μg/ml Poly d(A).[d]

Protocol 12. Hybridization of RNA filters[a]

All reagents should be RNase-free or DEPC-treated (*Protocol 7*).

Reagents

- 20 mM Tris–HCl, pH 8.0 (Tris base reacts with DEPC so dilute an autoclaved stock with DEPC-treated water)
- hybridization buffer (see p. 163–5)

Method

1. Wet filters bound with glyoxylated RNA in 20 mM Tris–HCl, pH 8.0 and then place into 20 mM Tris–HCl, pH 8.0 heated to 95°C. Remove the heat and recover the filter when the solution reaches room temperature. This step is necessary to remove glyoxal adducts from the RNA. Wet filters from the formaldehyde gels in sterile DEPC-treated water and proceed with pre-hybridization.

2. Place the filter in a plastic bag and add pre-heated hybridization buffer using 20 ml/100 cm² of filter. Seal the bag having excluded any air bubbles.

3. Pre-hybridize for approximately 2 h at the hybridization temperature and periodically re-distribute the solution about the filter. Perform the incubation by immersing the filter secured between glass plates into a waterbath, or use a commercially available hybridization oven.

4. Cut a corner off the bag, squeeze out the pre-hybridization mixture and replace it with hybridization mixture containing the probe denatured by heating in a boiling waterbath for 10 min (Section 5.2.2). Add 2.5 ml hybridization solution/100 cm². Exclude air bubbles, seal the bag, and hybridize.

[a] Protocol based largely on that given in ref. 24.

5.2.2 Probe choice
A range of probe types are available for the investigation of nucleic acid species. RNA blots may be probed using genomic DNA, cDNA, oligonucleotides (short, long, degenerate, singly, or in pools), PCR-synthesized probes,

[d] Use tRNA for end-labelled oligoprobes, and poly(A)/poly d(A) (Boehringer–Mannheim) for 3′ tailed probes (25).

or RNA probes. Choice is a matter of accessibility (many are commercially available), sequence knowledge, and aims of the experiment. Different types of probe require specific hybridization conditions.

i. Hybridization with genomic, cDNA, or PCR synthesized probes

The amount of probe used and hybridization period chosen depends largely on target abundance. Typically, hybridization is performed at 42°C for 16 h with the probe at a concentration of between 10–200 ng/ml of hybridization solution. It is important to note that the higher the probe concentration the greater the risk of background staining.

ii. Hybridization with oligonucleotide probes

Oligonucleotide probes are sometimes favoured because they are target-specific and there is no possibility of an erroneous signal from a labelled vector. They also give low background staining. Oligonucleotides covalently labelled with alkaline phosphatase are reportedly 10 times more sensitive than ^{32}P-labelled oligos (26). However blots containing > 30 µg of total cellular or > 5 µg of poly (A)+RNA are required because of the shortness of the target sequence.

The hybridization temperature is of key importance as short probes are more easily destabilized by high temperatures and low ionic strength than are conventional DNA probes. These probes do not usually require denaturation. Hybridization conditions need to be altered when using oligonucleotide probes, for example optimum hybridization time is often a matter of hours, and temperature needs to be calculated. Formamide is not included in the pre-hybridization and hybridization solutions, and carrier tRNA may be used to replace sonicated DNA to avoid possible quenching of the hybridization signal (p. 164). A probe concentration of 1–10 pmol/ml hybridization solution is suitable for these sequences.

iii. Hybridization with RNA probes

Stringent conditions are required when using RNA probes because of the stability of the RNA–RNA hybrids. The optimum hybridization temperature is typically between 55–68°C. Use between 100–200 ng of labelled RNA/ml hybridization solution and hybridize for over 16 hours.

5.2.3 Post-hybridization washes

Post-hybridization washing of filters is described in *Protocol 13*.

Protocol 13. Post-hybridization washing

Reagents

- 2 × SSPE: 0.36 M NaCl, 20 mM NaH$_2$PO$_4$, 2 mM EDTA, pH 7.2
- wash solution 1: 2 × SSPE, 0.1% (w/v) SDS
- wash solution 2: 0.1 × SSPE, 0.1% (w/v) SDS

Protocol 13. *Continued*

A. *DNA and RNA probe hybridized filters*

1. Wash at room temperature in two changes of wash solution 1 with > 50 ml/100 cm² filter for 5 min each.

2. Wash at 68°C in two changes of wash solution 2 for 15 min each.

B. *Oligonucleotide probes*

1. Wash the filters in two changes of wash solution 1 at room temperature for 5 min each.

2. Wash in wash solution 1 at the hybridization temperature for 15 min.

6. Detection

There are a number of commercial kits available for the detection of non-isotopic labels which are reviewed in Section 8. We present here protocols that are used routinely in our laboratories for colourimetric and chemiluminescent systems.

6.1 Detection of biotin and DIG-labelled probes

6.1.1 Biotinylated probes

Streptavidin (or avidin), which has a high affinity and specificity for biotin is covalently linked to the reporter enzyme (see Section 6.2) and then conjugated to the biotinylated probe. The enzyme catalyses a colour or chemiluminescent reaction (see Section 6.3). A second system, referred to as the 'sandwich technique' uses the multiple biotin binding sites on streptavidin, so that it is sandwiched between a biotinylated probe and a biotinylated enzyme (see *Protocol 14*).

6.1.2 Digoxigenin-labelled probes

Detection of DIG-labelled probes is by polyclonal anti-digoxigenin Fab-fragments conjugated to alkaline phosphatase, which are reported to exhibit less non-specific binding than complete antibodies. The end-point relies on an enzymatic detection with either a coloured precipitate or a light reaction. Although the binding constant for the DIG-hapten and DIG antibody is less than the biotin/streptavidin system, detection of 0.1 pg of homologous DNA after Southern transfer has been reported (4) (see *Protocol 15*).

6.2 Reporter enzymes

A number of reporter enzymes are available to catalyse both colourimetric and chemiluminescent reactions. Alkaline phosphatase (AP), which exhibits a high thermal stability, is the most widely used enzyme in both direct and indirect reactions. It is used in colourimetric reactions (Section 6.3.1), and several chemiluminescent compounds act as substrates for this enzyme, such as the modified dioxetanes (Section 6.3.2) and luciferin derivatives. Several types of luciferase enzymes, such as firefly luciferase from *Photanus pyralis*, form the basis of bioluminescent reactions (Section 6.3.2), and horseradish peroxidase (27) or hydrogen peroxidase catalyse chemiluminescent and colorimetric reactions in both direct and indirect systems.

Protocol 14 describes the detection reaction for biotinylated probes with alkaline phosphatase as the reporter enzyme.

Protocol 14. Detection of biotinylated probes

Equipment and Reagents

- buffer 1: 100 mM Tris–HCl, pH 7.5, 100 mM NaCl, 5 mM $MgCl_2$,[a] 0.05% Tween-20
- buffer 2: 1% blocking reagent,[a] 0.5% Tween-20[b] in buffer 1
- buffer 3: 100 mM Tris–HCl, pH 9.5, 100 mM NaCl, 20 mM $MgCl_2$

- Streptavidin/alkaline phosphatase conjugate (Gibco–BRL, Vector) or streptavidin and biotinylated alkaline phosphatase[c] (Gibco–BRL, Vector, Oncor).
- shaking table

Method

1. Following the post-hybridization washes, rinse the filters briefly in buffer 1.

2. Wash the filters vigorously in buffer 2 at room temperature for 1–1.5 h.

3. *Either* apply streptavidin–alkaline phosphatase conjugate (1 µg/ml) in buffer 1 and incubate for 15 min at room temperature with gentle rocking (use 3–4 ml/100 cm^2 of filter)
   or apply streptavidin (2 µg/ml) in buffer 2 for 10 min at room temperature, with gentle rocking. Wash with three changes of buffer 1 for 5 min each, then apply biotinylated alkaline phosphatase (1 µg/ml) in buffer 1 for 10 min, with gentle rocking. Use 3–4 ml/100 cm^2 of filter. The concentration of the conjugate may be titrated (Section 7.5).

4. Wash the filter in three changes of buffer 1 (50 ml/100 cm^2) for 5 min each.

5. Rinse briefly in buffer 3.

6. Proceed to *Protocol 16* for the colourimetric substrate or *Protocol 17* for the chemiluminescent substrate.

[a] Solutions for RNA filters should be treated as described in *Protocol 7*. Blocking reagent should be bought from commercial sources for RNA blots (see p. 163) to avoid RNase contamination. Non-fat dried milk or casein is suitable for DNA filters.

[b] The use of Tween-20 is especially recommended if using nitrocellulose filters and colorometric detection (33). Alternatively, use a blocking buffer composed of 3% BSA (Sigma), 0.18 M NaCl, 0.1 M Tris–HCl, pH 7.5, 0.05% Triton-X-100.

[c] The use of a two-stage system sometimes helps to alleviate background problems and gives increased sensitivity.

Protocol 15 describes the detection of DIG-labelled probes with alkaline phosphatase as the reporter enzyme.

Protocol 15. Detection of DIG-labelled probes

Equipment and reagents

- antidigoxigenin–AP, Fab fragments (Boehringer–Mannheim)
- non-fat dried milk (supermarket own brand) or commercially available blocking agents with recommended buffers[a]
- buffer 1: 100 mM Tris–HCl/150 mM NaCl, pH 7.5
- buffer 2: buffer 1 + 0.3% Tween-20
- buffer 3: buffer 1 + 3–4% non-fat dried milk[a]
- buffer 4: 100 mM Tris–HCl, pH 9.5, 100 mM NaCl, 50 mM MgCl$_2$
- shaking table

Method

All steps should be carried out on a shaking table at room temperature. Hybridization ovens have been found not to be suitable for the detection steps. Enough liquid to cover the filters is sufficient for Steps 1, 2, 3, 4, and 6. The antibody should be washed off in an excess of washing buffer with vigorous shaking.

1. Following post-hybridization washes, incubate the membrane briefly in buffer 2.
2. Incubate for 30 min in buffer 3.
3. Dilute anti-DIG alkaline phosphatase in buffer 1, 75 mU/ml for Hybond-N and 35 mU/ml for Nytran. Other filters will require titration, see Section 7.5. The antibody may be re-used several times within 24 hours if stored at 4°C.
4. Incubate the membrane for 30 min in the diluted antibody. Ensure the filter is covered and not allowed to dry out at any stage.
5. Wash the membrane in two changes of buffer 2 with vigorous shaking, for 2 × 15 min each.
6. Incubate the membrane for 5 min in buffer 4.
7. Proceed to *Protocol 16* for the colorimetric substrate or *Protocol 17* for the chemiluminescent substrate.

[a] See *Protocol 14* for blocking reagents for RNA blots.

6.3 Visualization techniques

6.3.1 Colourimetry

The colourimetric detection of non-radioactive probes is achieved by an enzyme-linked immunoassay with a colour reaction as the end-point. The hapten-labelled probe is hybridized to the DNA sequence and detected with a

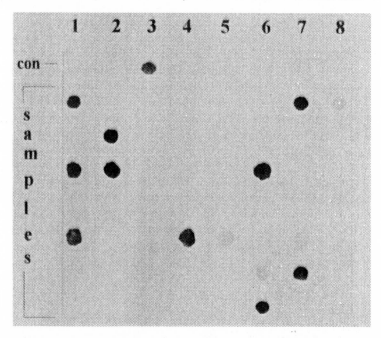

Figure 2. Dot blot colourimetric detection of HPV 16 DNA sequences from cytological samples. Cervical cells recovered from smear spatulae were digested and PCR-amplified using generic primers for the human papillomavirus (HPV) L1 region (36). A dilution containing 1 μl of product was spotted on to nitrocellulose filter (Gelman Biotrace) using a commercial apparatus. Following hybridization with PCR-synthesized biotinylated L1 sequence specific to the HPV type 16 (36, 37), the filter was washed in 0.1 × SSC, 0.1% SDS at 65 °C. Blocking and detection were carried out as described in *Protocols 14* and *16* with a 3 h incubation in NBT/BCIP. Control row 'Con' consists, from left to right, of L1 sequences amplified from HPV types 6, 11, 16, 18, 31, 33, and 35 (36) (positions marked by the numbers 1–7). The eighth sample is from a known HPV-negative sample. The cervical samples are represented in the bracketed 'sample' rows. Parallel blots were hybridized for the presence of other HPV types.

specific binding protein (such as avidin/streptavidin, anti-digoxigenin) conjugated to an enzyme. The most common colour system involves an alkaline phosphatase-mediated reaction with 5-bromo-4-chloro-3-indolyl phosphate (BCIP) and Nitroblue tetrazolium salt (NBT), e.g. see *Figure 2*. The BCIP undergoes oxidation to release a phosphoryl group, whilst the NBT is reduced to diformazan resulting in the formation of a blue or brown precipitate adhering to the membrane. *Protocol 16* describes the colourimetric detection of biotin or DIG-labelled probes.

6.3.2 Chemiluminescence detection

Chemiluminescence (CL) is defined as the chemical production of light. These systems also involve an enzyme-linked immunoassay with a light reaction as the end-point. Where the luminescent compounds are found in biological systems the term bioluminescence (BL) is used. Many CL systems utilize the high energy bonds in the dioxetanes. Most dioxetanes are unstable, but substituents, such as adamantyl, have been found to confer stability and permit the synthesis of a dioxetane that can be activated by enzymes. AMPPD^R, ($C_{18}H_{21}O_7PNa_2$), 3-(4-methoxyspiro{1,2-dioxetane-3,2'-tricyclo-[3.3.1.1$^{3.7}$]-decan}4-yl)phenyl phosphate, is an example of a modified dioxetane with a half-life of a year, and is used as a direct chemiluminescent substrate for phosphatases (28). Enzymatic dephosphorylation by alkaline phosphatase results in the oxidation of a high energy bond and the accumulation of a partially stable intermediate, before decomposition into adamantone and methyl meta-oxybenzoate which emits light at 477 nm over several days. The image is captured on X-ray film. *Protocol 17* describes the detection of biotin or DIG-labelled probes by chemiluminescence (*Figures 3* and *4*).

Protocol 16. Colourimetric detection

Equipment and reagents

- Buffer 3: 100 mM Tris–HCl, pH 9.5, 100 mM NaCl, 20 mM MgCl$_2$
- Nitroblue tetrazolium (NBT): 75 mg/ml in 70% dimethylformamide (DMF)
- 5-bromo-chloro-3-indolyl phosphate (BCIP) (Sigma): 50 mg/ml in DMF
- plastic bags

Method

1. Continue from Steps 6 or 7 in *Protocol 14* or *15*.

2. Prepare substrate solution by adding 44 μl of NBT and 35 μl of BCIP to 10 ml of buffer 3. Shake the buffer whilst adding the chromogens and filter through a 0.2 μm membrane to help avoid background staining.

3. Place filters in a plastic bag, add 10 ml of substrate solution/100 cm^2 filter, seal, and store in the dark. Filters may be left for 12–16 h or until the desired signal is apparent.

4. Stop the reaction by rinsing the filter with copious volumes of water. Replacing the substrate solution after 3–4 h helps to improve the sensitivity and keeps the background levels low.

5. Filters may be photocopied, sealed in their bags during the course of incubation, or wet after the reaction has been stopped. Signal and any background fades when the filter dries, but is restored by re-wetting.

6. Store the filters sealed in plastic bags and in the dark.

Protocol 17. Chemiluminescent detection

Equipment and reagents

- *either* Luminogen PPD (Boehringer–Mannheim) plus buffer 4: (100 mM Tris–HCl, pH 9.5, 100 mM NaCl, 50 mM MgCl$_2$[a])
- *or* AMPPD (Tropix-NBL) plus reaction buffer 0.1 M dietholamine, 1 mM MgCl$_2$, 0.02% sodium azide buffer, pH 10.

- strong plastic sheets
- Whatmann 3MM paper
- plastic bags
- X-ray film

Method

1. Continue from Steps 6 or 7 in *Protocol 14* or *15*.

2. Dilute chemiluminescent substrate in reaction buffer (final concentration of 0.25 mM or 0.125 mM depending on membrane — see Step 3, *Protocol 15*).

3. Place the membrane, DNA side up, on a strong piece of plastic. Place dilute substrate on the membrane and spread gently with a second piece of plastic. Approximately 1 ml will cover 100 cm^2. Leave for 2 min then blot gently on a sheet of Whatmann 3MM paper. Do not allow to dry.

4. Seal in plastic bags, incubate for 30–60 min at 37°C then expose to Kodak-X-Omat or any standard X-ray film at room temperature. Optimum exposure time depends on the type of probe, strength of signal, and level of non-specific background. For unique sequence probes we expose the film for 3 h initially, and then expose the membrane to a second sheet of film overnight if necessary. Multiple exposures are possible as the luminescence increases for several days.

5. For re-probing the membrane must be kept damp (Section 6.4). Store at −20°C to avoid bacterial growth.

[a] Mg^{2+} ions are not essential (Section 7.3) but a pH above 9.5 is critical for the substrate reaction.

Other CL compounds are available, such as the cyclic diacylhydrazides, for example luminol (27). In one system (Amersham International — see Section 8.3) the probe is labelled with fluorescein-11-dUTP and detected by an enzyme-linked anti-fluorescein antibody. Horseradish peroxidase catalyses the oxidation of luminol to form a chemically-excited form of 3-aminophthalate dianion. As this returns to ground-state, light is emitted at 428 nm for a very short time. Enhancers increase and prolong the light output to allow it to be captured on film.

Biotinylated probes may also be detected by a streptavidin/glucose-6-phosphate dehydrogenase conjugate instead of alkaline phosphatase. The

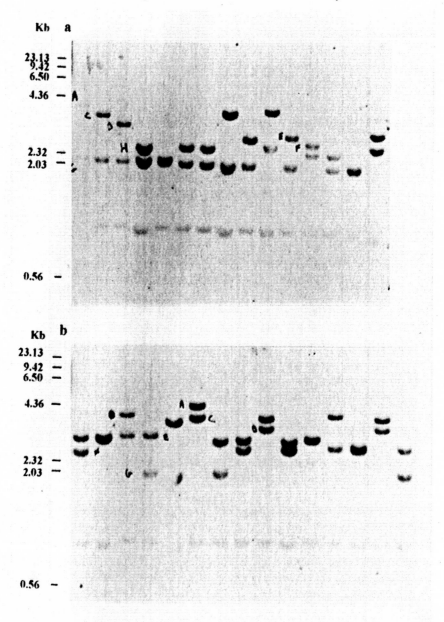

luminescent reaction is catalysed by flavine mononucleotide (FMN) oxido-reductase and luciferase (29). Luciferin derivatives may also be used in a BL reaction with alkaline phosphatase and luciferase. The luciferase/luciferin BL system and hydrogen peroxide CL techniques are reviewed extensively in refs 30 and 31.

Figure 3. (a) Various human genomic DNA samples were digested with *Hin*fl, electrophoresed on an 0.8% agarose gel (approx. 5 μg per lane), and Southern blotted. The Southern blot was probed with a single copy genomic clone MS205 (5.2 kb) labelled with DIG–dUTP. DNA bands which hybridized with the probe were detected as described in *Protocols 15* and *17*. The X-ray film was developed after 4 h. (Reproduced with permission from Dr P. C. Harris.) (b) Various human genomic DNA samples were digested with *Hin*fl, electrophoresed on an 0.8% agarose gel (approx. 5 μg per lane), and Southern blotted. The Southern blot was probed with a single copy genomic clone MS205 (5.2 kb) labelled with DIG–dUTP. DNA bands which hybridized with the probe were detected as described in *Protocols 15* and *17*. The X-ray film was developed after 4 h. (Reproduced with permission from Dr P. C. Harris.)

6.3.3 Fluoresence

Fluoresence-based systems are beset with problems from non-specific background fluorescence present in all biological samples. This is overcome with the use of time-resolved fluoresence which requires specific instrumentation to detect the fluorescence. A protocol for the detection of alkaline phosphatase labels conjugated to biotinylated or DIG-labelled probes by time-resolved fluorescence is described in ref. 32.

6.3.4 Direct detection

A direct assay system was developed (34) to avoid the need for 'sandwich amplification' in the detector system. As described in Chapter 1 the probes are covalently linked to the reporter enzyme and, after hybridization, react

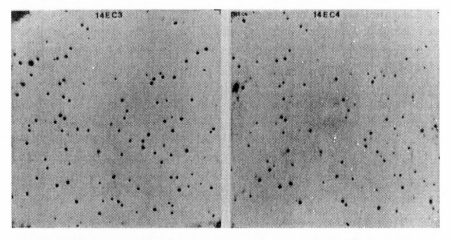

Figure 4. Phage library constructed from YAC 14E, with duplicate filters (14EC3 and 14EC4) made by standard techniques. 100–200 ng of DIG-labelled human DNA in 20 ml formamide was hybridized and detected as described in *Protocols 15* and *17* with overnight exposure.

directly with the colour or chemiluminescent substrate. The system reported in ref. 34 was optimized for a colour reaction achieved with either hydrogen peroxidase or alkaline phosphatase, or the two reactions sequentially on the same filter to produce two colours on one filter with different probes. To detect probes conjugated directly to alkaline phosphatase, follow *Protocols 16* and *17*.

6.4 Quantitation of Dot blots

Filters that have been detected by CL methods and have a final result in the form of an autoradiograph can be analysed in the same way as filters hybridized with radiolabelled probes. Densitometry allows a comparison of the density of the bands on the X-ray film with known standards hybridized with the sample to be assessed. For accurate quantitation the film must not be overexposed as the signal may exceed the linear response of the film. The CL technique described in *Protocol 17* allows several exposures to be made. Filter membranes with a colour precipitate may also be analysed by densitometric methods (35). Luminometers are commercially available which are designed for the direct measurement of the signal produced in a CL reaction from the filter rather than the X-ray film. This method may be a cheaper option than if a densitometer is not available.

6.5 Stripping membranes

6.5.1 Stripping membranes after a colour reaction

Stripping a probe from filters after a colour reaction requires harsh treatment and is possible only with nylon membranes. The colour precipitate is removed by incubation in dimethylformamide at 50–60°C. This procedure must be carried out in a fume hood. The probe is then removed from the membrane by standard procedures.

6.5.2 Stripping probes after a chemiluminescence reaction

For filters hybridized with biotinylated probes:

(a) Agitate the membrane in $1 \times$ SSC/1% SDS heated to 95°C for 5 min.

(b) Wash in two changes of $1 \times$ SSC at room temperature for 15 min each, allow to air-dry, or store damp at $- 20$°C.

For filters hybridized with DIG-labelled probes:

(a) Incubate in 0.2 M NaOH/0.1%(w/v) SDS at 37°C for 2×15 min.

(b) Rinse in $2 \times$ SSC.

In both cases the efficiency of stripping is tested by re-detection with the appropriate protocol. If a signal is re-detected, the stripping protocol can be repeated until no signal is produced. However, the efficiency of both of these stripping protocols does vary with the membrane used. It is essential that the filters are never allowed to dry out otherwise it will not be possible to remov the probe and it will be re-detected in subsequent reactions. Another protoco based on proteinase K treatment followed by incubation in hot formamide has been recommended for DIG-labelled probes (25).

6.6 Sensitivity

For non-radioactive methods to be a viable alternative to radiolabelled hybridizations they must give a sensitivity equal to or surpassing that of radio-activity. This is dependent on the hybridization kinetics, total signal, and signal to noise ratio being at least equivalent to that of radiolabelled probes.

Colourimetric methods, being one of the original non-radioactive methods, do not routinely attain the required levels of sensitivity needed to detect many single copy sequences, having an optimized limit of 100 fg for Southern blots, although detection of 50 fg has been reported on Dot blots (32). However, some sequences require a sensitivity of low femtogram levels. The arrival of chemiluminescent substrates has provided a system whose signal is at least equivalent to ^{32}P-labelled probes (30 fg), and which allows the detection of single copy genes routinely on genomic blots both on standard and pulsed-field gels (*Figure 5*). Whilst some problems do occur with the variability of signal and efficiency of stripping on some batches of membrane, further optimization of the noise/signal ratio may produce a greater potential sensitivity

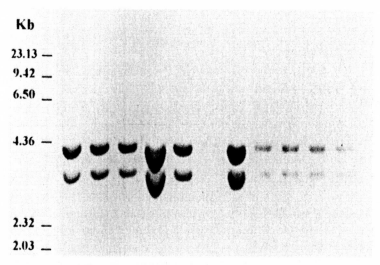

Figure 5. Various human genomic DNA samples were digested with *Sau* 3A, electro-phoresed on an 0.8% agarose gel (approx. 5 μg per lane), and Southern blotted. The Southern blot was probed with a single copy genomic clone PJE (0.5 kb) labelled with DIG–dUTP. DNA bands which hybridized with the probe were detected as described in *Protocols 15* and *17*. X-ray film was developed after 4 h. (Reproduced with permission from Dr P. C. Harris.)

than traditional radiolabelled methods, detected in a significantly shorter time.

7. Troubleshooting

7.1 Low sensitivity

If the results show a clear background but a weak signal, several factors may be altered to increase the sensitivity of the system.

(a) Check the labelling efficiency of the probe (see *Protocol 18* and *Protocol 3* footnote *d*, for RNA probes).

(b) Increase the concentration of probe in the hybridization mix.

(c) Check that the concentration of the conjugate and substrate are optimized (see Section 7.5).

(d) Check that the pH of the substrate buffer is above pH 9.5.

For chemiluminescence:

(e) Increase the exposure time of the filter to the film.

(f) Increase the time of the 37°C incubation.

(g) If all conditions are optimized, rinse the filter in buffer 3 or 4 and add fresh substrate.

For colourimetry:

(h) Incubate the filter for up to 16 h and change the NBT/BCIP solution after 3–4 h.

7.2 High background

(a) Increase the stringency washes after hybridization. If this is thought to be the problem, re-wash the filter at a higher stringency and re-detect the probe. It will not be necessary to re-hybridize the filter.

(b) There may be too much probe in the hybridization mixture.

(c) Insufficient blocking during either the hybridization or the detection steps. Check that there is sonicated salmon sperm DNA in the hybridization mixture or add/increase the concentration of a commercial blocking agent. Increase the concentration and time of incubation of the blocking solution in the detection.

(d) The concentration of the conjugate and substrate is too high for the type of membrane (see *Protocol 19*).

(e) Increase the washing steps after conjugate incubation.

(f) Filter-sterilize the NBT/BCIP substrate solution through a 0.2 μm filter.

7.3 Spots on the X-ray film

(a) Bacterial or fungal contamination may be a problem on old filters that have been stored damp. If the filter is not to be stripped wash in $2 \times SSC$ and store at $-20\,°C$.

(b) There may be a precipitate in the probe solution. Ensure the hybridization mixture is completely dissolved before use. Remove the Mg^{2+} ions from the substrate buffers. Filter the conjugate through a 0.45 μm membrane.

7.4 Labelling efficiency

One of the disadvantages of the non-radioactive system is the difficulty in assessing the labelling efficiency. The most direct method is to Dot blot the purified probe on to a membrane and quantify it (see *Protocol 18*).

Protocol 18. Assessment of labelling efficiency

Equipment and reagents

- purified probe
- dH₂O
- blotting membrane
- detection reagents (see *Protocols 14–17*)

Method

1. Serially dilute the purified probe in water to produce a range of probe concentrations, for example 2 ng–0.2 pg.

2. Spot 1 μl of each dilution on to the membrane of choice and UV treat or bake.

3. Detect the probe as described in *Protocols 14–17*.

Whilst the results are difficult to convert to a percentage incorporation, they do give an indication of low incorporation as a potential problem. For a more accurate result it is possible to compare the dilutions of the labelled probe to known standards (Boehringer–Mannheim) spotted on to the same filter. This is worth doing for probes in frequent use that have been generated in bulk.

7.5 Titration of conjugate and luminogen substrate

The degree of signal to background is dependent not only on the probe concentration in the hybridization mixture but also the concentration of the conjugate and substrate as well. It is advisable to titrate the conjugate to find the optimal concentration for a good signal to background ratio (see *Protocol 19*). This is particularly important when testing new batches or brand of membrane.

Protocol 19. Titration of conjugate and substrate

Equipment and reagents

- labelled DNA
- membrane
- detection reagents (see *Protocols 14–17*)

Method

1. Produce a number of identical blots with the labelled DNA.
2. Detect all together as *Protocol 14* or *15* until the conjugate step.
3. Make a range of dilutions of the conjugate [a] in the relevant buffer, using the suppliers recommendation as a guide (e.g. if 1/10 000 is recommended, try 1/2500, 1/5000, 1/10 1000, 1/15 000, and 1/20 000).
4. Incubate separately in conjugate.
5. Return to detection protocols.
6. After detection assess results for the best signal to background ratio.

[a] The luminescent substrate may also be titrated by keeping the conjugate concentration constant and varying the substrate concentration.

7.6 Enhancers

Luminescence amplifying equipment and reagents (LAMs) (Tropix) enhance light emission from chemiluminescent substrates by providing a hydrophobic environment to prevent quenching of the signal. Nitro-Block is essential if a chemiluminescent substrate is used with a nitrocellulose filter. Additional enhancers, such as Emerald, have also been used successfully to enhance the signal, but should not usually be required.

8. Commercial kits

There are a number of commercial kits available for the non-isotopic detection of nucleic acids which form a good starting place to working with these techniques. We list several of them here without bias or comment as to their efficiency. We suggest that the relevant literature for the kits is read thoroughly before choosing a specific one to try.

8.1 Labelling kits

See Section 3.1.1 for standard labelling reactions. There are several kits available for direct labelling with alkaline phosphatase:

- Cambridge Research Biochemicals — E-link system

- Gibco–BRL – ACES Labelling system,
- Promega – LIGHTSMITH and Quantum Yield

8.2 Detection systems

8.2.1 Biotinylated probes detected with the NBT/BCIP colour reaction

The following kits utilize the biotin/streptavidin/alkaline phosphatase reaction in various combinations as described in Section 6:

- Gibco–BRL – BluGENE non-radioactive Nucleic Acid Detection System
- Oncor – Sure Blot Hybridization and Detection kit
- Sigma – Cool Probe

8.2.2 Biotinylated probes detected with a chemiluminescence reaction

The following kits utilize biotin/streptavidin/AP with chemiluminescence as described in Section 6:

- Gibco–BRL – Photogene Nucleic Acid detection system
- Millipore – Polar Plex Chemiluminescent Blotting Kit
- New England Biolabs – Phototape
- Promega – Quantum Yield Chemiluminescence
- USB – Gene Images Kit

8.2.3 DIG-labelled probes detected with a colour reaction

- Boehringer – Mannheim – Nucleic Acid Detection Kit

8.2.4 DIG-labelled probes detected with a chemiluminescence reaction

- Boehringer – Mannheim – DIG Luminescent Detection Kit

8.3 Enhanced chemiluminescence (ECL) from Amersham International Ltd

This system is summarized in Section 6.3.2. Amersham International Ltd also produce a direct labelling system where the positively charged horseradish peroxidase complex is bound electrostatically to the negatively charged nucleic acid. Hybridization takes place in a specially designed gold buffer to stabilize the enzyme.

8.4. Others

8.4.1 Bio-Rad – Gene-Lite Chemiluminescent Detection Kit

This involves the primary hybridization of single-stranded phagemid or M13 DNA containing the probe DNA to a blot, followed by a secondary hybridization with an oligonucleotide–alkaline phosphatase conjugate complementary to the universal primer present in the vector sequence. The alkaline phosphatase catalyses a light reaction as described in Section 6. This system avoids the need to label the probe sequence.

8.4.2 Multicolour Detection Set from Boehringer–Mannheim

This allows nucleic acid sequences to be detected with different coloured, ('rainbow detection'), hybridization signals on the same blot. Sequences are labelled with digoxigenin, fluorescein, or biotin and hybridized simultaneously to the membrane. The probes are detected with the respective antibody or streptavidin–alkaline phosphatase conjugate. Each label is visualized sequentially by immunoassay with different combinations of naphtholophosphate and diazonium salt to give one of three colours directly on the membrane.

Acknowledgements

We would like to thank Peter C. Harris and Sandra Thomas of the MRC Molecular Haematology Unit for constructive comments on the manuscript and for the use of their chemiluminescent autoradiographs.

References

1. Kricka, L. J. (ed.) (1992). *Nonisotopic DNA probe techniques*. Academic Press, London.
2. Langer, P. R., Waldrop, A. A., and Ward, D. C. (1981). *Proc. Natl Acad. Sci. USA*, **78**, 6633.
3. Leary, J. J., Brigati, D. J., and Ward, D. C. (1983). *Proc. Natl Acad. Sci. USA*, **80**, 4045.
4. Kessler, C., Holtke, H.-J., Seible, R., Burg, J., and Muhlegger, K. (1990). *Biol. Chem. Hoppe-Seyler*, **371**, 917.
5. Rigby, P. W. J., Dieckmann, M., Rhodes, C., and Berg, P. (1977). *J. Mol. Biol.*, **113**, 237.
6. Feinberg, A. P. and Vogelstein, B. (1983). *Anal. Biochem.*, **132**, 6.
7. Tarleton, J. and Schwartz, C. E. (1991). *Clin. Genet.*, **39**, 121.
8. Lo, Y.-M. D., Mehal, W. Z., and Fleming, K. A. (1988). *Nucleic Acids Res.*, **16**, 8719.
9. Seible, R., Holtke, H.-J., Ruger, R., Meindl, A., Zachau, H. G., Rabhofer, R., Roggendorf, M., Wolf, H., Arnold, N., Weinberg, J., and Kessler, C. (1990). *Biol. Chem. Hoppe-Seyler*, **371**, 939.
10. Kessler, C. (1992). In *Nonisotopic DNA probe techniques* (ed. L. J. Kricka), pp. 29–92. Academic Press, London.
11. Stickland, J. (1992). In *Diagnostic molecular pathology: a practical approach*, Volume 2 (ed. C. S. Herrington and J. O'D. McGee), pp. 25–64. IRL Press, Oxford.

12. Sambrook, J., Fritsch, E. F., and Maniatis, T. (1989). *Molecular cloning: a laboratory* manual (2nd edn). Cold Spring Harbor Laboratory Press, Cold Spring Harbor, New York.
13. Smith, C. L., Klco, S. R., and Cantor, C. R. (1988). In *Genome analysis: a practical approach* (ed. K. E. Davies), pp. 41–72. IRL Press, Oxford.
14. Jeyaseelan, K. (1987). In *Genes and proteins: a laboratory manual of selected techniques in molecular biology* (ed. K. Jeyaseelan, M. C. M. Chung, and O. L. Kon), pp. 17–24. ICSU Press, Paris.
15. Southern, E. M. (1975). *J. Mol. Biol.*, **98**, 503.
16. Allefs, L. H. M., Salentijn, E. M. J., Krens, F. A., and Rouwendal, G. J. A. (1990). *Nucleic Acids Res.*, **18**, 3099.
17. Farrel R. E. Jr. (1993). *RNA methodologies: a laboratory guide for isolation and characterisation*. Academic Press, London.
18. Thomas, P. S. (1983). In *Methods in Enzymology*, Vol. 100 (ed. R. Wu), p. 255. Academic Press, New York.
19. Lehrach, H., Diamond, D., Wozeny, J. M., and Boedtker, H. (1977). *Biochemistry*, **16**, 4743.
20. Alwine, J. C., Kemp, D. J., Paker, B. A., Reiser, J., Renaart, J., Stark, G. R., and Wahl, G. M. (1979). *In Methods in Enzymology*, Vol. 68 (ed. R. Wu), p. 220. Academic Press, New York.
21. McMaster, G. K. and Carmichael, G. C. (1977). *Proc. Natl Acad. Sci. USA*, **74**, 4835.
22. Hames, B. D. and Higgins, S. J. (ed.) (1985). *Nucleic acid hybridization: a practical approach*. IRL Press, Oxford.
23. Wahl, G. M., Berger, S. L., and Kimmel, A. R. (1987). In *Guide to molecular cloning* (ed. S. L. Berger and A. R. Kimmel), pp. 399–406. Academic Press, London.
24. Holtke, H.-J. and Kessler, C. (1990). *Nucleic Acid Res.*, **18**, 5943.
25. Boehringer–Mannheim (1993). *The Dig system users' guide for filter hybridization*. Boehringer–Mannheim, Germany.
26. Gibco–BRL Life Technologies (1993). *New products bulletin*, No. 11, Issue 1. Gibco-BRL Life Technologies, P.O. Box 35, Trident House, Paisley, Scotland.
27. Stone, T. and Durrant, I. (1991). *GATA*, **8**, 230.
28. Tizard, R., Cate, R. L., Ramachandran, K. L., Wysk, M., Voyta, J. C., Murphy, O. J., and Bronstein, I. (1990). *Proc. Natl Acad. Sci. USA*, **87**, 4514.
29. Nicolas, J.-C., Balaguer, P., Terouanne, B., Villebrun, M. A., and Boussioux, A.-M. (1992). In *Nonisotopic DNA probe techniques* (ed. L. J. Kricka), pp. 203–25. Academic Press, London.
30. Kricka, L. J. (1988). *Anal. Biochem.*, **175**, 14.
31. Whitehead, T. P., Kricka, L. J., Carter, T. J. N., and Thorpe, G. H. G. (1979). *Clin. Chem.*, **25**, 1531.
32. Templeton, E. G., Wong, H. E., and Pollak, A. (1992). In *Nonisotopic DNA probe techniques* (ed. L. J. Kricka), pp. 95–111. Academic Press, London.
33. Chan, V. T.-W., Fleming, K. A., and McGee, J.O'D. (1985). *Nucl. Acids Res.*, **13**, 8083.
34. Renz, M. and Kurz, C. (1984). *Nucl. Acids Res.*, **12**, 3435.
35. Lee, A. S. G. and McGee, J. O'D. (1989). *Nucl. Acid. Res.*, **17**, 2364.
36. Bauer, H. M., Greer, C. E., and Manos, M. M. (1992). In *Diagnostic molecular*

pathology; a practical approach, Volume 2 (ed C. S. Herrington and J. O'D McGee), pp. 131–52. IRL Press, Oxford.
37. Lo, D. Y.-M., Mehal, W. Z., and Fleming, K. A. (1990). In *PCR protocols; a guide to methods and applications* (ed. M. A. Innis, D. H. Gelfand, J. J. Sninsky, and T. J. White), pp. 113–18. Academic Press, London.

7

Non-isotopic DNA analysis

Y.-M. DENNIS LO and WAJAHAT Z. MEHAL

1. Introduction

Advances in molecular biology have resulted in a rapid increase in knowledge of the human genome. To date, a large number of polymorphic systems has been described, many of which, for example the human leukocyte antigen (HLA) system, have many alleles and play an important role in human disease. The accurate and rapid genotyping of these multi-allelic systems, therefore, has important applications in research and potentially clinical diagnosis.

The advent of the polymerase chain reaction (PCR) (1) has provided us with a powerful method for amplifying DNA sequences. The enormous amplification factor offered by the PCR allows the visualization of amplified products non-isotopically. PCR-based approaches for genotyping can generally be divided into two broad categories:

(a) those in which the PCR serves as the amplification step, with allelic determination carried out by a subsequent process, and

(b) those in which the PCR step serves as both the amplification and diagnostic steps.

In this chapter, we shall concentrate on three PCR-based strategies for genotyping multi-allelic systems:

(a) the amplification refractory mutation system (ARMS)

(b) artificial restriction fragment length polymorphism (A-RFLP) analysis

(c) single-strand conformation polymorphism (SSCP) analysis.

2. Amplification refractory mutation system (ARMS)

2.1 Principles

ARMS or allele-specific PCR (2) is an amplification strategy in which a primer is designed in such a way that it is able to discriminate between templates which differ in a single nucleotide residue. In other words, an

ARMS primer can be designed to amplify a specific member of a multi-allelic system whilst remaining refractory to amplification of another allele which may differ for as little as a single base from the former. ARMS is based on the principle that the *Taq* polymerase, the DNA polymerase commonly used in PCR, lacks 3' to 5' exonuclease activity, and thus a mismatch between the 3' end of the PCR primers and the template will result in greatly reduced amplification efficiency (*Figure 1*). Thus, an ARMS typing system can be designed by constructing primers with their 3' nucleotide overlying the polymorphic residue. Hence, one ARMS primer can be constructed to amplify specifically one allele of a multi-allelic system. For typing a system with *n* alleles, *n* ARMS primers will be required with the typing achieved in *n* reactions.

Recently, two variants of the ARMS theme have been described: multiplex ARMS (3) and double ARMS (4). Multiplex ARMS was developed to facilitate the application of ARMS to genotyping multi-allelic systems: a number of ARMS primers, each specific for a particular allele, are present in a single reaction in a single tube. PCR products corresponding to different alleles can then be distinguished by physical characteristics such as length.

In double ARMS, two allele-specific ARMS primers are used simultaneously in a single reaction. As each ARMS primer hybridizes at a different polymorphic site, double ARMS allows direct haplotype determination without the need for pedigree analysis and without resorting to single molecule dilution (5), or single sperm typing (6). It is also very useful for genotyping multi-allelic systems in which individual alleles are distinguished from each other by having different combinations of polymorphic residues, as a positive signal is only present when the right combination of ARMS primers is used. Another advantage of double ARMS is that it has greatly enhanced specificity compared with single ARMS. This may prove useful in the detection of a minority DNA population amongst a background of related but non-identical DNA molecules, for example for the study of the phenomenon of chimaerism following bone marrow transplantation.

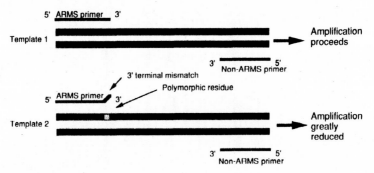

Figure 1. Principles of ARMS. A primer mismatched at its 3' end to the template will result in greatly reduced amplification efficiency.

2.2 Design of ARMS primers

Primer design is the most important aspect in creating a working ARMS-based typing system. The ARMS concept requires that the nucleotide or nucleotides distinguishing the various alleles be placed at the 3′ end of the ARMS primer. Different authors have slightly different views as to the type of mismatches which are most discriminatory for ARMS analysis (2, 7, 8): in our hands, A–G or G–A mismatches are best. For other types of nucleotide mismatch, we routinely introduce an artificial mismatch at the residue 1 or 2 bases from the 3′ end of the primer to enhance the specificity of the primer further (2). Primers ranging from 14 to 30 bases have been used for ARMS analysis (2, 7, 8). Relatively long primers of 30 bases have been advocated for this purpose (2), and we have found that primers of this length work well. For a given sequence, an ARMS primer can be constructed to prime in the sense or the anti-sense direction. Thus, if a given ARMS primer is found to be non-discriminatory in one direction, it is worthwhile constructing another one in the opposite direction.

2.3 Optimization of ARMS reactions

The most important optimization parameter for allele-specificity is the annealing temperature. We routinely start our optimization with an annealing temperature of 55°C and vary the temperature in 3°C steps. Though some authors emphasize the importance of optimizing $MgCl_2$ concentration, we find that a concentration of 1.5 mM works well for many of our applications. The 'hot start' technique is recommended by some (2, 9) for ARMS analysis. In our experience, this modification is most useful for certain 'problematic' primers; for most primer–template combinations, however, conventional non-hot start PCR appears to be adequate. dNTP concentrations have also been found to affect ARMS specificity, with a low dNTP concentration resulting in a more specific reaction, though sometimes with reduced sensitivity (8). We find that a dNTP concentration of 100 μM for each of dATP, dCTP, dGTP, and TTP works well for most ARMS primers.

As the absence of amplification in an ARMS reaction is used to exclude the presence of a specific allele, most authors include an internal positive control to guard against amplification failures. In our laboratory, we routinely optimize the conditions for an ARMS primer without the internal control primers. The latter are introduced after the parameters for ARMS analysis have been established.

2.4 Practical considerations for multiplex ARMS

As discussed above, multiplex ARMS was developed in order to simplify the analysis of multi-allelic systems so that fewer reactions are required. All the primers for multiplex ARMS should be designed so that they exhibit allele-

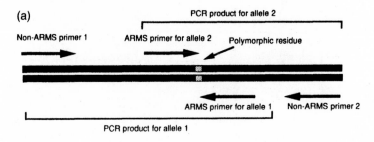

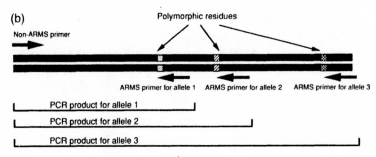

Figure 2. Examples of multiplex ARMS formats. (a) A configuration set up for a two-allelic system. Note that apart from the ARMS PCR products for allele 1 and allele 2, there is also an internal positive control product formed by the two non-ARMS primers. The two ARMS products should be designed to be of different lengths. See Powis *et al.* (21) for an example of this system in application. (b) A configuration set-up for a three-allelic system. Note that the products of the different alleles are distinguishable by size. See *Figure 3* for an example of this configuration.

specificity at the same annealing temperature. Furthermore, if different primers have different amplification efficiencies, these have to be balanced by adjusting the respective primer concentrations (3).

A number of formats for multiplex ARMS are potentially possible. Two of these configurations are illustrated in *Figure 2*.

2.5 DNA samples for ARMS analysis

As ARMS analysis depends on the relative, rather than the absolute difference in efficiency for amplifying a matched or mismatched template, there exists a 'window' of specificity. In other words, an ARMS primer which exhibits allele-specificity at 25 cycles may no longer be specific at 40 cycles of PCR. Consequently, the amount of DNA template used should be relatively constant for reproducible ARMS-based typing. As a guideline, it is reasonable to optimize ARMS conditions using 1 µg genomic DNA and 30 cycles of PCR. For sample sources in which the yield of amplifiable DNA is variable, e.g. paraffin wax embedded archival materials, ARMS analysis may not be as robust as some of the other genotyping methods mentioned later in this chapter. If ARMS is intended to be used for DNA extracted from these sources, a two-step procedure may be employed: an initial 'amplification phase', aiming at amplifying a specific DNA segment from the material, and a second 'diagnostic phase', in which a predetermined amount of the first phase PCR product is used for ARMS analysis (10).

2.6 Procedures

As described above, the design aspects of an ARMS-based system are the most important part for setting up a working assay. Once the design and optimization of amplification conditions have been achieved, the execution of ARMS genotyping is relatively straightforward, analagous to that of conventional PCR (see *Protocol 1*).

Protocol 1. ARMS analysis of genomic DNA

Reagents

- 10 × PCR mixture: 100 mM Tris–HCl, pH 8.3, 500 mM KCl, 15 mM MgCl$_2$, 0.1% gelatin
- PCR primers: including an ARMS primer (or a combination of ARMS primers for multiplex ARMS), a downstream non-ARMS primer and internal control primers[a]
- dNTPs: typically 100 μM each of dATP, dCTP, dGTP, and TTP

- genomic DNA
- sterile distilled water
- *Taq* polymerase
- mineral oil
- 1.5% agarose gel stained with ethidium bromide

Method

1. Prepare PCR mixture as follows:
   - 10 × PCR reaction mixture   10 μl
   - PCR primers   x μl
   - dNTPs   y μl
   - genomic DNA   1 μg
   - sterile distilled water (SDW)   to 100 μl total volume
   - *Taq* polymerase   2.5 Units

2. Overlay with 100 μl of mineral oil.

3. Denature at 94°C for 8 min.

4. Carry out thermal cycling, e.g. 30 cycles of thermal denaturation at 94°C for 1 min, primer annealing at 55°C for 1 min, and extension at 72°C for 1 min.

5. Perform a final incubation at 72°C for 8 min following the last PCR cycle.

Protocol 1. *Continued*

6. Analyse 10 μl of the reaction on a 1.5% agarose gel stained with ethidium bromide.

ᵃ Primer concentration is determined empirically, but is typically 100 pmol per 100 μl reaction volume.

Figure 3 illustrates the results of a multiplex ARMS analysis designed to distinguish three groups of alleles at the HLA *DRB3* locus (3). The three groups of alleles are HLA DRw52a, b, and c. ARMS products are designed to be distinguishable from each other on the basis of size.

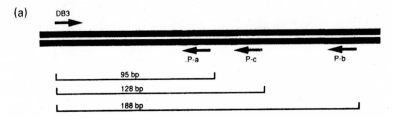

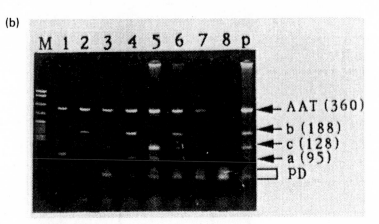

Figure 3. Multiplex ARMS for HLA typing. (a) Relative locations of the primers are shown. P-a, P-b, and P-c are ARMS primers designed to specifically amplify DRw52a, b, and c, respectively. DB3 is an upstream non-ARMS primer. Primer sequences are listed in *Table 1*. (b) The results of multiplex ARMS typing are shown. Lane 1, PCR product from an individual with DRw52a; lane 2, DRw52b; lane 3, DRw52c; lane 4, DRw52a/b heterozygote; lane 5, DRw52a/c heterozygote; lane 6, DRw52b/c heterozygote; lane 7, no *DRB3* alleles; lane 8, negative control (water). Lane M, pBR322 DNA digested with *MspI* (marker). Lane p, PCR product ladder. AAT, a, b, and c indicate positions of alpha-1-antitrypsin positive control, PCR products from the *DRB3*0101* (DRw52a), *DRB3*0201/2* (DRw52b), and *DRB3*0301* (DRw52c), respectively. Numbers in parentheses indicate sizes of PCR products in bp. PD denotes primer dimers. (*Figure 3b* reproduced from Lo *et al.* (1991). *Lancet*, **338**, 65, with permission.)

Table 1. Sequences of PCR primers

Primer name	Sequence
DB3	5'-GACCACGTTTCTTGGAGCT-3'
P-a	5'-CTCCTGGTTATGGAAGTATCTGTCCACGT-3'
P-b	5'-GTCCTTCTGGCTGTTCCAGTACTCGGAAT-3'
P-c	5'-CTCCCCCACGTCGCTGTCGAAGCGCACGG-3'
AAT-1	5'-CCCACCTTCCCCTCTCTCCAGGCAAATGGG-3'
AAT-2	5'-GGGCCTCAGTCCCAACATGGCTAAGAGGTG-3'
P-1	5'-TTCCTTCTGGCTGTTCCAGTACTCGGAG-3'
P-2	5'-TTCCTTCTGGCTGTTCCAGTACTCGGAA-3'
DB130	5'-AGGGATCCCCGCAGAGGATTTCGTGTACC-3'
COMM-52	5'-CTGCTCCAGGAGGTCCTTCTGGCTGTTCCA-3'
938	5'-AAT GGA TGA TTT GAT GCT GTC CC-3'
939'	5'-TCT GGG AAG GGA CAG AAG ATG AC-3'

3. Artificial restriction fragment length polymorphism (A-RFLP) analysis

3.1 Principles

Restriction analysis of PCR products was one of the earliest techniques used for analysing amplification products (11). This approach is applicable to the distinction of alleles in which the polymorphic residue results in the creation or removal of a restriction enzyme site. Unfortunately, many polymorphisms are not associated with restriction enzyme site changes and thus are not amenable to this analysis. However, by site-directed mutagenesis using primers with mismatches near the 3' ends, it is possible to create an artificial RFLP (A-RFLP) for almost all naturally occurring DNA polymorphisms (12, 13). *Figure 4* illustrates the principles of this approach.

3.2 Design of primers for A-RFLP

An A-RFLP primer can be designed easily and rapidly using a semi-automatic approach using a computer program which will search for restriction enzyme sites for a given sequence, for example DNA Strider. The process is illustrated in *Figure 5*. The assumptions are that the polymorphic residue is P and that we are searching for restriction enzymes with recognition sites of up to six bases. Five bases on either side of P are entered into the computer (from −5 to +5) and the program is used to search for a restriction enzyme site encompassing P. If a restriction enzyme site is found which is only present in one allele but not in the other one, then no further searching is required. If no restriction site polymorphism is found then the nucleotides from −2 to −5 and from +2 to +5 are changed one at a time, with a computer search being carried out after each alteration. For each position, the nucleotides A, T, C, and G are substituted in turn (one of them will be found in the naturally

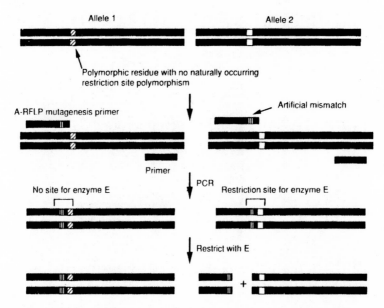

Figure 4. Principles of artificial RFLP. Note that the mutagenesis primer has its 3′ end just before the polymorphic site, as compared with ARMS when it is overlying the polymorphic site.

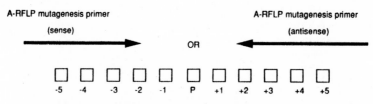

Figure 5. Design of A-RFLP primers. See text for details.

occurring sequence). We avoid changing the −1 or the +1 position (which will be used as the last base of the PCR primer) as this may reduce the amplification efficiency (see Section 2.1). It is worthwhile carrying on the process until all the possibilities have been investigated, as more than one solution may be possible for a given polymorphism and some restriction enzymes work better than others.

Following endonuclease restriction, the PCR product from the allele with the restriction site will have the portion containing the A-RFLP primer cleaved off, thus resulting in a smaller sized fragment on agarose gel electrophoresis. We typically use A-RFLP primers which are 30 bases in length. The optimal size of the PCR product is around 120 bp. Agarose gels of 1.5% normally give adequate resolution for allelic discrimination.

3.3 Procedures

Allelic assignment by A-RFLP is a two-step process, with an initial amplifica-tion step, followed by a diagnostic restriction step (see *Protocol 2*).

Protocol 2. Genotyping by A-RFLP

Reagents

- PCR mixture (see *Protocol 1*)
- primers
- dNTPs
- sterile distilled water (SDW)

- restriction enzyme
- 10 × restriction buffer
- 1.5% agarose gel

Method

1. Set up the PCR as detailed for *Protocol 1*, substituting the appropriate primers.

2. Perform thermal cycling as determined empirically to achieve specificity (see Section 2).

3. Set up restriction mixture consisting of:
   - PCR product   10 µl
   - SDW   7 µl
   - 10 × restriction buffer   2 µl
   - restriction enzyme [a]   1 µl

4. Restrict for 2 h at the incubation temperature for the enzyme used.

5. Analyse 10 µl of the restricted PCR product in a 1.5% agarose gel stained with ethidium bromide. [b]

[a] Concentrated preparations of restriction enzymes (> 40 µ/ml) have been found to give the best results.
[b] Include a lane containing the unrestricted PCR product for comparison.

Figure 6 shows an example of A-RFLP analysis of the HLA-*DQB1* locus. The 17 alleles at this locus could be divided into those coding for aspartic acid at codon 57 and those which do not. The aspartic acid status at codon 57 is important for determining susceptibility and resistance to insulin-dependent diabetes mellitus (14). The various *DQB1* alleles coding for aspartic acid at position 57 have, in common, GA as the first two nucleotides of the codon (*Figure 6a*). Codon 58 is GCC for every allele. Thus, a *Hin*fI site (GANTC) diagnostic for the aspartate-57 alleles can be created by mutating codon 58 from GCC to TCC via appropriately constructed mutagenesis primers. As the third base of codon 57 could be C or T, two PCR primers are necessary for direct amplification of these two groups of alleles. By using this A-RFLP system, the 17 alleles can be divided into those with aspartate-57 and those

(a)

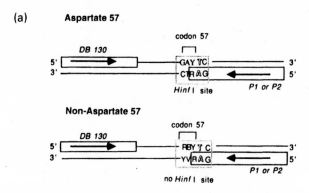

(b)

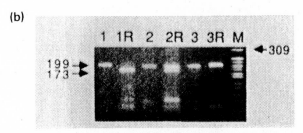

Figure 6. Typing of aspartate-57 status at the *DQB1* locus by A-RFLP. (a) Diagram illustrating the principle of A-RFLP detection of aspartate-57 and non-aspartate-57 *DQB1* alleles. The 199 bp PCR product is shown. P1 and P2 are the mutagenesis primers. DB130 is an upstream primer common to all *DQB1* alleles. Note the penultimate nucleotide A of primers P1 and P2 is an artificial mismatch. Y = T or C; R = A or G; B = T, G, or C; V = A, C, or G (IUPAC nomenclature). (b) 1, 2, and 3 represent undigested PCR products from three individuals. 1R, 2R, and 3R represent the PCR products digested with *Hin*fI. Individual 1 = aspartate-57 homozygote; individual 2 = aspartate-57/non-aspartate-57 heterozygote; and individual 3 = non-aspartate-57 homozygote. Lane M, pBR322 DNA digested with *Msp*I (marker). 199 = uncut PCR product in bp, 173 = cut PCR product in bp, 309 = 309 bp band of the size marker. (*Figures 6a, b* reproduced with permission of Springer–Verlag from: Patel *et al.* (1992). *Immunogenetics*, **36**, 264.)

without aspartate-57 in a simple system using PCR followed by restriction with the enzyme *Hin*fI (15).

4. Single-strand conformation polymorphism (SSCP) analysis

4.1 Principles

Single-strand conformation polymorphism (SSCP) analysis is based on the principle that the mobility of single-stranded DNA in non-denaturing polyacrylamide gels is dependent not only on its length but also on its sequence (16). It is thought that this phenomenon is due to secondary conformations taken up by single-stranded DNA. SSCP analysis has been used to distinguish alleles which differ by as little as a single base. Unlike ARMS and A-RFLP, SSCP analysis does not require prior knowledge of the sequence of the alleles and thus is useful in detecting previously unknown alleles or mutations.

SSCP analysis is useful for PCR products of up to 400 bp in length. Longer PCR products may be cleaved into smaller fragments before being subjected to SSCP analysis. Electrophoretic conditions (acrylamide concentration, glycerol content, running temperature, and composition of electrophoretic buffer) have an important influence on the mobility of single-stranded DNA. Unfortunately, there are no established formulae for predicting the optimal conditions for detecting a particular allele, and the choice of electrophoretic conditions is essentially a trial-and-error process. In general, we find two conditions to be most useful:

(a) 6% polyacrylamide gel with 5% glycerol in $1 \times$ Tris–borate–EDTA (TBE) at 18°C

(b) 6% polyacrylamide gel without glycerol in $1 \times$ TBE at 4°C.

To aid the interpretation of SSCP gels, we routinely include an undenatured sample to demonstrate the position of double-stranded DNA. It should also be noted that single-stranded DNA may take up more than one conformation. Thus, it is possible to have more than two single-strand bands for a homozygous individual.

4.2 Procedures

The amplification phase of PCR-SSCP analysis is essentially the same as for conventional PCR (17). There are three methods by which single-stranded DNA can be detected non-isotopically: ethidium bromide staining (ref. 17 and see *Protocol 3*), silver staining (ref. 18 and see *Protocol 4*), and fluorescence (ref. 19 and 20 and see *Protocol 5*).

Protocol 3. SSCP analysis using ethidium bromide staining

Reagents

- acrylamide
- 10 × TBE: 0.9 M Tris base, 0.9 M boric acid, 20 mM EDTA, adjust the pH to 8.1–8.2 using solid boric acid
- 10% (w/v) ammonium persulfate

- TEMED: *N,N,N′,N′*-tetramethylenediamine
- glycerol
- 0.5% bromophenol blue, 0.5% xylene cyanol in deionized formamide
- sterile distilled water (SDW)

Method

1. Set up PCR as detailed in *Protocol 1* and ref 17.

2. Perform thermal cycling with empirically derived parameters to achieve specificity.

Protocol 3. *Continued*

3. Set up a polyacrylamide gel as follows [a]:
   - 40% acrylamide [b]   12.5 ml
   - 10 × TBE   5 ml
   - 10% ammonium persulfate   320 μl
   - TEMED   40 μl
   - glycerol [c]   2.5 ml
   - SDW   to 50 ml final volume

4. Mix 6–30 μl of PCR product [d] with 1/10 volume of 0.5% bromophenol blue/0.5% xylene–cyanol in deionized formamide.

5. Denature at 85°C for 10 min.

6. Chill on ice for 5 min.

7. Load polyacrylamide gel. [e]

8. Electrophorese at chosen temperature [f] at 10–30 V/cm for 1.5–3 h.

9. Stain in ethidium bromide (0.5 mg/ml) for 10–40 min.

10. Visualize and photograph under UV transillumination.

[a] We use a Bio-Rad Protean II electrophoresis apparatus.
[b] Concentration of gel could be varied empirically to give optimal results.
[c] Glycerol is normally omitted for gels run at 4°C.
[d] 1 pmol or more per track gives good results. The detection limit is 0.3 pmol.
[e] An undenatured control lane should be included for comparison purposes.
[f] The temperature of electrophoresis should be carefully controlled. We use a water-cooled jacket connected to a thermostat for temperature control.

Figure 7 (adapted from ref. 22) illustrates an example of SSCP analysis using ethidium bromide staining as the detection method. The single-base G to C substitution at position 72 in exon 4 of the p53 gene was analysed by PCR amplification of a 185 bp fragment and SSCP analysis. A constant amount of MDA cell line DNA (homozygous for the G allele) was present in each sample, with a variable amount of plasmid DNA (C allele). The slower migrating single-stranded (ss) DNA demonstrated a sequence-dependent mobility shift, which allowed the G and C alleles to be differentiated.

Protocol 4. SSCP analysis by silver staining

Equipment and reagents

- PCR equipment and reagents (see *Protocol 1*)
- polyacrylamide gel (see *Protocol 3*)
- 0.5% bromophenol blue, 0.5% xylene–cyanol in deionized formamide
- silver staining kit (Bio-Rad)
- gel drier

Method

1. Set up PCR as detailed in *Protocol 1*.
2. Perform thermal cycling.
3. Set up a polyacrylamide gel (see *Protocol 3*).
4. Mix 6–30 μl of PCR product [a] with 1/10 volume of 0.5% bromophenol blue/0.5% xylene–cyanol in deionized formamide.
5. Denature at 85°C for 10 min.
6. Chill on ice for 5 min.
7. Load the polyacrylamide gel.
8. Electrophorese at chosen temperature at 15–30 V/cm for 1.5–3 h.
9. Stain using a silver staining kit (Bio-Rad). [b]
10. Dry in a gel drier at 80°C for 45 min.
11. Document the results by photography or photocopying.

[a] We find that approximately 1 pmol of PCR product gives good results, though smaller amounts are also detectable.

[b] Water purity is extremely important for silver staining. Use clean glassware and gloves (previously rinsed in SDW) for the whole process.

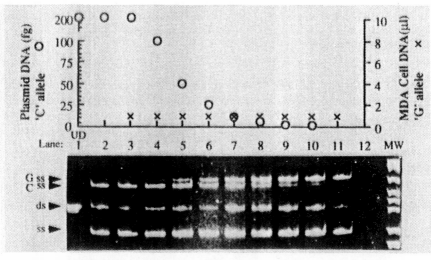

Figure 7. SSCP analysis by ethidium bromide staining. A serial dilution of plasmid DNA containing the C allele was mixed with a constant amount of MDA cell DNA containing the G allele, as illustrated in the graph. This mixed DNA population was amplified with primers 938 and 939' (see *Table 1*), and the products run on a denaturing SSCP gel stained with ethidium bromide. The single-stranded (ss) and double-stranded (ds) DNA of the two alleles C and G were easily resolved. The slower migrating ssDNA demonstrated a sequence specific mobility shift. Lane 1, undenatured (UD) DNA; lane 2, plasmid DNA (C allele); lanes 3–10, titration of plasmid DNA with constant amount of MDA cell DNA (allele G); lane 11, MDA cell DNA; lane 12, blank control; lane MW, pBR322 DNA digested with *Msp*I (marker).

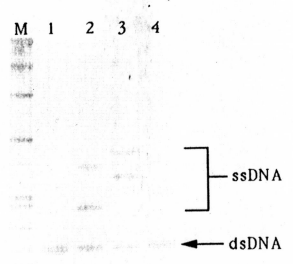

Figure 8. SSCP analysis by silver staining. Genomic DNA was amplified using the primers DB3 and COMM-52 (see *Table 1*) and analysed by SSCP. Lane M: pBR322 DNA digested with *Mspl* (marker); lane 1, undenatured PCR product (200 bp); lanes 2–4, SSCP analysis; lane 2, DRw52a homozygote; lane 3, DRw52b homozygote; lane 4, DRw52c homozygote. Positions of double-stranded DNA (dsDNA) and single-stranded DNA (ssDNA) are marked. (Reproduced by permission of Oxford University Press: from Lo *et al.* (1992). *Nucl. Acids Res.*, **20**, 1005.)

Figure 8 shows the use of SSCP for typing the three groups of alleles at the HLA *DRB3* locus which codes for DRw52a, b, and c. As illustrated in the figure, the three groups of alleles are easily distinguishable by SSCP. We found that heterozygous individuals, for example a DRw52a/b individual, apart from giving rise to four bands corresponding to single-strand DNA, also had characteristic heteroduplexes migrating just above the double-stranded DNA.

Protocol 5. SSCP analysis using fluorescence primers

Equipment and reagents
- PCR equipment and reagents (see *Protocol 1*)
- fluorescence-labelled primers[a]
- ethanol
- sterile distilled water (SDW)

- 0.5% bromophenol blue, 0.5% xylene–cyanol in deionized formamide
- polyacrylamide gel (see *Protocol 3*)
- UV transilluminator

Method

1. Set up the PCR as detailed in *Protocol 1*.

2. Perform thermal cycling using empirically derived parameters (see Section 2.5).

3. Transfer 1 µl of the PCR product into a second round of asymmetric PCR using 30 pmol of a fluorescence-labelled primer.[a]

4. Perform 30 cycles of asymmetric PCR.

5. Ethanol precipitate the whole of the second round product.

6. Resuspend in 20 µl SDW.

7. Add 2 µl of 0.5% bromophenol blue/0.5% xylene–cyanol in deionized formamide.

8. Heat at 85°C for 10 min.

9. Chill on ice for 5 min.

10. Load polyacrylamide gel.

11. Electrophorese at chosen temperature at 15–30 V/cm for 1.5–3 h.

12. Visualize by UV transillumination.[b]

13. Mask the area outside the area of interest for improved contrast.

14. Photograph the gel.[c]

[a] Fluorescence labels made by different companies may perform differently. In our laboratory we use 5′-carboxyfluorescein (FAM) and 6-carboxy-X-rhodamine (ROX) labelled primers synthesized by Applied Biosystems (20).
[b] For colour photography UV transilluminators from different manufacturers may require different filtration. In our laboratory we use a 302 nm transilluminator made by UVP.
[c] For black and white photography, a rhodamine-labelled primer is preferred to a fluorescein-labelled one, as conventional Polaroid systems set up for photographing ethidium bromide stained gels are applicable. For reference, a 25A filter with Polaroid 667 film (ISO 3000) will require an exposure of 2 sec at f/4.5. For colour photography, Polaroid 668 color film (ISO 80), a Wratten No. 8, and a 2B filter are required. Exposure is at f/4.5 for 5 min.

Figure 9 shows a black and white photograph of fluorescence SSCP using a rhodamine-labelled primer. The system is used to distinguish the alleles coding for HLA DRw52a, b, and c. Note that a single band is seen for a homozygous individual as only one single strand is produced by the asymmetric PCR step. For an example of colour photography of fluorescence SSCP results, see ref. 19.

4.3 Comparison of non-isotopic SSCP techniques

Of the three non-isotopic SSCP systems mentioned above, ethidium bromide staining is the most time-efficient. However, we find that it is not as sensitive and the results not as reproducible as those obtained using silver staining.

In our laboratory, silver staining is the preferred method for SSCP analysis. Though relatively time-consuming, with the staining procedure taking up to 3 h, the results are clean and reproducible, with better sensitivity than both ethidium bromide based and fluorescence primer based technologies. The stained and dried gel can be scrutinized for subtle mobility shifts which may easily be missed on a smaller than life-size photograph of an ethidium-stained gel.

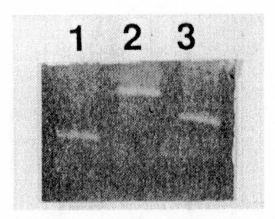

Figure 9. Fluorescence SSCP analysis of *DRB3* locus. Genomic DNA was amplified using the primers DB3 and COMM-52 (see *Table 1*). A second round asymmetric PCR using a ROX-labelled DB3 primer was used as described in *Protocol 5*. Lane 1, amplified product from a homozygous DRw52a individual; lane 2, amplified product from a homozygous DRw52b individual; and lane 3, amplified product from a homozygous DRw52c individual. Exposure was 2 sec at *f*/4.5 using Polaroid 667 film (ISO 3000).

Fluorescence SSCP is still at a relatively early stage of development. The relatively low sensitivity offered by a single label on a PCR primer demands that a large amount of PCR product be loaded per track. However, at this high concentration of DNA, single-stranded DNA molecules reanneal very rapidly. In order to overcome this limitation, we perform asymmetric PCR to produce large quantities of single-stranded DNA products which will not be hybridized to molecules from the complementary strand. We envisage that the full potential of fluorescence SSCP will be realized when temperature-controlled fluorescence scanning devices become commercially available at a reasonable price. The provision of a reference marker labelled with a different coloured label per track could allow rapid and reliable allelic assignment (19).

5. Choice of method for genotyping

We have outlined three generally applicable methods for the analysis of multi-allelic systems. Of these methods, ARMS analysis is the most time-efficient, as both the amplification and diagnostic steps are combined together. With the introduction of multiplex ARMS the system is very efficient for analysing loci with a large number of alleles. However, the advantage of ARMS is also the cause of its disadvantage, as robustness may be compromised. For example, a lack of amplification, which would be taken as indicating the absence of a particular allele, may simply be the result of PCR failure. The inclusion of an internal positive control only partly solves the problem as it is theoretically possible for an ARMS reaction to fail despite having apparently working positive controls.

A-RFLP is probably the most robust of the three methods. It is also very efficient in detecting heterozygous individuals. As the amplification and diagnostic restriction steps are separated, there is always a control for potential PCR failures. Thus, we recommend its use for DNA sources in which the yield of amplifiable materials may be variable, for example paraffin wax embedded archival material.

SSCP analysis is probably the most time-consuming of the three methods as the polyacrylamide gel electrophoresis and the subsequent detection step, for example silver staining, both require time. However, it can potentially detect previously unknown alleles or mutations. Furthermore, it is relatively easy to set up SSCP analysis for a multi-allelic series as a minimum of only one set of PCR primers will be needed. For ARMS or A-RFLP analysis, however, multiple primer sets will be required.

A very powerful way of using these methods is to combine them. One example is to use ARMS for most of the alleles in a system but to incorporate A-RFLP for a proportion of the alleles. The robustness of the latter is then used to cross-check the accuracy of the more time-efficient but less reliable ARMS typing. Another way of combining these systems would be to sub-divide a locus with a large number of alleles into a number of subgroups using ARMS, followed by typing of subgroups using SSCP (19). Potentially, a large number of variations on this theme is possible.

6. Conclusions

In this chapter, we have described three methods for genotyping multi-allelic systems. The combinations of these methods should be adequate for most genotyping situations. A number of other genotyping methods have been tried in our laboratory, but we find those described here to be the most easily implemented.

Acknowledgement

The authors are Wellcome Medical Graduate Fellows.

References

1. Saiki, R. K., Gelfand, D. H., Stoffel, S., Scharf, S. J., Higuchi, R., Horn, G. T., Mullis, K. B., and Erlich, H. A. (1988). *Science*, **239**, 487.
2. Newton, C. R., Graham, A., Hepstinstall, L. E., Powell, S. J., Summers, C., Kalsheker, N., Smith, J. C., and Markham, A. F. (1989). *Nucl. Acids Res.*, **17**, 2503.

3. Lo, Y.-M. D., Mehal, W. Z., Wordsworth, B. P., Chapman, R. W., Fleming, K. A., Bell, J. I., and Wainscoat, J. S. (1991). *Lancet*, **338**, 65.
4. Lo, Y.-M. D., Patel, P., Newton, C. R., Markham, A. F., Fleming, K. A., and Wainscoat, J. S. (1991). *Nucl. Acids Res.*, **19**, 3561.
5. Ruano, G., Kidd, K. K., and Stephens, J. C. (1990). *Proc. Natl Acad. Sci. USA*, **87**, 6296.
6. Li, H., Cui, X., and Arnheim, N. (1990). *Proc. Natl Acad. Sci. USA*, **87**, 4580.
7. Wu, D. Y., Ugozzoli, L., Pol, B. K., and Wallace, R. B. (1989). *Proc. Natl Acad. Sci. USA*, **86**, 2759.
8. Kwok, S., Kellogg, D. E., McKinney, N., Spasic, D., Goda, L., Levenson, C., and Sninsky, J. J. (1990). *Nucl. Acids Res.*, **18**, 999.
9. Chou, Q., Russell, M., Birch, D. E., Raymond, J., and Bloch, W. (1992). *Nucl. Acids Res.*, **20**, 1717.
10. Lo, E.-S. F., Lo, Y.-M. D., Tse, C. H., and Fleming, K. A. (1992). *J Clin. Pathol.*, **45**, 689.
11. Saiki, R. K., Scharf, S., Faloona, F., Mullis, K. B., Horn, G. T., Erlich, H. A., and Arnheim, N. (1985). *Science*, **230**, 1350.
12. Haliassos, A., Chomel, J. C., Tesson, L., Baudis, M., Kruh, J., Kaplan, J. C., and Kitzis, A. (1989). *Nucl. Acids Res.*, **17**, 3606.
13. Eiken, H. G., Odland, E., Boman, H., Skjelkvak L., Engebretsen, L. F., and Apold, J. (1991). *Nucl. Acids Res.*, **19**, 1427.
14. Todd, J. A., Bell, J. I., and McDevitt, H. O. (1987). *Nature*, **329**, 599.
15. Patel, P., Lo, Y.-M. D., Bell, J. I., and Wainscoat, J. S. (1992). *Immunogenetics*, **36**, 264.
16. Orita, M., Suzuki, Y., Seikiya, T., and Hayashi, K. (1989). *Genomics*, **5**, 874.
17. Yap, E. P. H. and McGee, J. O'D. (1992). *Trends Genet.*, **8**, 489.
18. Ainsworth, P. J., Surh, L. C., and Coulter-Mackie, M. B. (1991). *Nucl. Acids Res.*, **19**, 405.
19. Lo, Y.-M. D., Patel, P., Mehal, W. Z., Fleming, K. A., Bell, J. I., and Wainscoat, J. S. (1992). *Nucl. Acids Res.*, **20**, 1005.
20. Chehab, F. F. and Kan, Y. W. (1989). *Proc. Natl Acad. Sci. USA*, **86**, 9178.
21. Powis, S. H., Tonks, S., Mockridge, I., Kelly, A. P., Bodmer, J. G., and Trowsdale, J. (1993). *Immunogenetics*, **37**, 373.
22. Yap, E. P. H. and McGee, J. O'D. (1993). In *PCR technology: current innovations* (ed. H. G. Griffin and A. M. Griffin). CRC Press, Florida.

8

PCR analysis of RNA

HIROSHI YOKOZAKI and EIICHI TAHARA

1. Introduction

The polymerase chain reaction (PCR) has become an invaluable technique for amplifying small amounts of DNA for a wide range of molecular biological analyses. First strand cDNA, which is the product of reverse transcription of messenger RNA, can also serve as a template for PCR, such that gene expression can be analysed from only trace amounts of mRNA. It can be applied not only to the molecular cloning of genes, but also to RNA blot analysis (1), nuclease protection assay (2), and mRNA phenotyping (3) for the study of low copy number transcripts.

2. Reverse transcriptase based techniques for RNA analysis

The enzyme reverse transcriptase (RT) synthesizes first strand cDNA using mRNA as template. By combining this reaction with subsequent polymerase chain reaction (PCR) amplification, low abundance mRNA species can be analysed in cells or tissue specimens. In this section, conventional protocols for the extraction of total RNA, the RT reaction, and for PCR after the RT reaction are given.

It is important to note that all procedures involving RNA must be free from ribonuclease (RNase) contamination. The following procedure should be applied to all glassware, plasticware, and solutions:

(a) Soak all the glassware to be used in the manipulation of RNA in water containing 0.1% (w/v) diethyl pyrocarbonate (DEPC) for 2 h, rinse several times with autoclaved water, then autoclave for 30 min at 15 p.s.i. (120°C).

(b) Use sterile disposable plasticware, which is essentially free of RNases.

(c) Water should be treated with 0.1% DEPC for at least 12 h and autoclaved for 30 min at 15 p.s.i. (120°C) on a liquid cycle. Prepare all buffers and solutions with DEPC-treated water and using RNase free reagents.

The major source of RNase contamination is the worker's hand. Disposable latex or plastic gloves should, therefore, always be worn during all procedures.

2.1 Extraction of total RNA

Tissues and cells should be snap-frozen in liquid nitrogen immediately after removal or harvest for the optimization of RNA yield and quality. Two conventional protocols for the extraction of total RNA from tissue samples or cultured cells are detailed here. Both protocols use guanidium isothiocyanate (GTC), a strong protein denaturant which can inactivate RNase, and are based on the method of Chirgwin *et al.* (4). *Protocol 1* requires ultracentrifugation and relatively large amounts of starting material, but the isolated RNA is extremely pure. On the other hand, only a high-speed centrifuge, or even a microcentrifuge, is required in *Protocol 2*, and the total RNA obtained is good enough for RT-PCR amplification.

Protocol 1. Extraction of total RNA from a relatively large sample

(*This protocol requires an ultracentrifuge*)

Reagents

- guanidium isothiocyanate mixture (GTC mixture): 4.85 M GTC, 70 mM sodium acetate, 0.5% sarcosyl, 1% 2-mercaptoethanol; add 2-mercaptoethanol just prior to use
- caesium chloride cushion: 5.7 M CsCl, 0.1 M EDTA; adjust the pH to between 6.5 and 7.0 with HCl

- TES buffer: 10 mM Tris–HCl, pH 7.4, 1 mM EDTA, 10% SDS
- chloroform: *n*-butanol (4:1)
- 5 M NaCl

Method

1. Crush frozen specimens with a hammer, then homogenize in a 50 ml polypropylene centrifuge tube (Falcon cat. no. 2070 or its equivalent) with 6 ml of GTC mixture using a Polytron mixer.

2. Transfer the homogenate to 13 ml thick-wall centrifuge tubes (Sarstedt cat. no. 60.540S or equivalent) and centrifuge at 12 000 *g* at 30°C for 30 min.

3. Transfer the supernatant to new Sarstedt tubes and add 2.6 g caesium chloride. Mix well.

4. Add 1.2 ml of CsCl cushion to each of two 13 × 51 mm ultracentrifuge tubes and slowly overlay half of the CsCl-mixed sample from Step 3 into each tube. Balance the tubes with GTC mixture.

5. Ultracentrifuge at 120 000 *g* at 20°C for more than 12 h using a Beckman SW50.1 rotor or its equivalent.

6. After ultracentrifugation, aspirate out about 80% of the supernatant and wash the wall of the tubes with GTC mixture, twice. Invert the

tubes and aspirate out all of the supernatant with a Pasteur pipette and air-dry the tubes. Check that the transparent RNA pellet is at the bottom of the centrifuge tube.

7. Cut off the bottom of the centrifuge tubes with a razor blade, then dissolve the RNA pellets in 1.5 ml of TES buffer

8. Transfer the dissolved RNA to 6 ml polypropylene centrifuge tubes (Sarstedt cat. no. 60.546S or equivalent) and add 1.5 ml of chloroform:n-butanol.

9. Mix well, centrifuge at 2000 g at 20°C for 15 min and transfer the upper layers to a 13 ml polypropylene centrifuge tube.

10. Add a further 1.5 ml TES buffer to the lower layers, mix well, and centrifuge at 2000 g at 20°C for 15 min. Combine the upper layers from Steps 9 and 10.

11. Add 150 µl of 5 M NaCl and 9 ml ethanol, mix well, and place the tube at −20°C overnight.

12. Centrifuge at 12 000 g at 4°C for 30 min. Discard the supernatant and invert the tube on a paper towel for 5 min to dry the pellet.

13. Add 500 µl of DEPC-treated H_2O and dissolve the RNA pellet. Add 597 µl of DEPC-treated water to 3 µl of RNA solution and measure the optical density (OD) at 260 nm. (8 × OD mg/ml gives the concentration of the RNA solution.)

14. Dilute the RNA solution with DEPC-treated water to 0.5–1.0 mg/ml and aliquot. Store at −80°C.

Protocol 2. Extraction of total RNA from small samples

(*This protocol does not require an ultracentrifuge*)

Equipment and reagents

- solution E: 4 M guanidium isothiocyanate (GTC), 25 mM sodium citrate, pH 7.0, 0.5% sarcosyl, 0.1 M 2-mercaptoethanol. To minimize handling, a stock solution is prepared by dissolving 250 g of GTC in the manufacturer's bottle with 293 ml of water, 17.6 ml of 0.75 M sodium citrate, pH 7.0, and 26.4 ml 10% sarcosyl at 65°C. This stock solution can be stored for 3 months at room temperature. Complete solution E is prepared adding 0.36 ml 2-mercaptoethanol for every 50 ml of stock solution.
- 2 M sodium acetate, pH 4.0
- water-saturated phenol
- chloroform:isoamyl alcohol mixture (49:1)
- isopropanol (ice-cold)
- 75% ethanol
- 0.5% SDS
- Seal-a-Meal bags

Method

1. Transfer the tissue into a Seal-a-Meal bag, previously cooled in dry-ice and pulverize it with a hammer, a few strokes at a time, putting it back on dry-ice each time.

Protocol 2. *Continued*

2. Transfer the tissue quickly into a baked glass homogenizer containing 1 ml of solution E for every 100 mg of tissue. Homogenize with several strokes or until the DNA is evidently sheared.[a]

3. Add 1/10 volume of 2 M sodium acetate (pH 4.0), and shake well.

4. Add 1 volume of water-saturated phenol and shake well.

5. Add 0.2 ml of chloroform:isoamyl alcohol mixture (49:1) for every 1 ml of initial solution E. Shake vigorously for 10 sec, then cool on ice for 10–15 min.

6. Centrifuge at 10 000 *g* for 20 min at 4°C.

7. Transfer the aqueous phase into a clean tube. Precipitate the nucleic acid by adding an equal volume of cold isopropanol and incubating at −20°C for 1 h. Centrifuge again as for Step 6.

8. Dissolve the pellet in 0.3 ml of solution E for every 100 mg of tissue. Precipitate again by repeating Steps 3–7.

9. Suspend the pellet in 75% ethanol. Centrifuge at 10 000 *g* for 20 min at 4°C and remove as much ethanol as possible. Dry the pellet and dissolve it in 0.5% SDS or other appropriate buffer (heat up to 65°C for 5 min if the pellet is difficult to dissolve). Measure the optical density at 260 nm and calculate the RNA concentration (see *Protocol 1*).

10. Aliquot and store the samples as described in *Protocol 1*.

[a] Samples can be homogenized in a microcentrifuge tube with a plastic pestle fitted to the shape of the tube.

2.2 Reverse transcription polymerase chain reaction (RT-PCR)

The RT-PCR has became a standard laboratory technique for the detection of gene expression. Although pre-mixed ready-to-use kits (e.g. GeneAmp RNA PCR kit from Cetus) are available for this method, the whole system can be constructed easily from individual components.

2.2.1 Reverse transcription

Protocol 3. Reverse transcription

Reagents

- solution A (10 × PCR buffer): 500 mM KCl, 100 mM Tris–HCl, pH 8.3
- solution B: 25 mM $MgCl_2$
- solution C: 50 mM random hexamer in 10 mM Tris–HCl, pH 8.3
- solution D (dNTP mix): 10 mM dATP, 10 mM dCTP, 10 mM dGTP and 10 mM dTTP in 10 mM Tris–HCl, pH 8.3

- Moloney murine leukaemia virus reverse transcriptase: 50 U/ml (Pharmacia–LKB)
- RNase inhibitor: 20 U/ml (Stratagene)
- DEPC-treated, H_2O
- total RNA

Method

1. Denature total RNA by heating at 65°C for 10 min.

2. Add the following, in order, to a microcentrifuge tube:

   - Solution A   2 µl
   - Solution B   4 µl
   - Solution C   1 µl
   - Solution D   8 µl
   - total RNA (denatured) x µl ($\leqslant$ 1 mg total RNA)
   - reverse transcriptase   1 µl
   - DEPC-treated H_2O to a final volume of 20 µl
                                       (in a 500 µl PCR tube)

3. Programme and run a thermal cycler for reverse transcription as follows:

   - 24°C   10 min
   - 42°C   40 min
   - 99°C   5 min
   - 4°C   5 min

2.2.2 Polymerase chain reaction amplification after reverse transcription

The RT reaction products can be used immediately, without changing the tube, for PCR amplification by simply adjusting the buffer and $MgCl_2$ concentration.

Protocol 4. Polymerase chain reaction

Reagents

- solution A (as in *Protocol 3*)
- solution B (as in *Protocol 3*)
- *Taq* DNA polymerase: 5 U/ml
- PCR primers (see Section 3)
- mineral oil
- chloroform

Methods

1. For each reverse transcription sample (20 µl) prepare a minimum of 78 µl of PCR reaction mix as follows:

   - solution A   8 µl
   - solution B   4 µl
   - sterile distilled water   66.5 µl
   - *Taq* DNA polymerase   0.5 µl
   - Total   78 µl

2. Dispense the PCR primers [a] into each tube; 1 µl each of the downstream and upstream primers.

Protocol 4. *Continued*

3. Overlay the reaction mixture with 50–100 µl mineral oil and start a thermal cycler programmed as follows:

94°C	1 min	1 cycle
94°C	1 min	
55°C	2 min	30 cycles
70°C	3 min	
70°C	10 min	1 cycle
4°C	soak	

4. Remove the mineral oil with chloroform. The PCR reaction products can be stored at 4°C or −20°C before further analysis.

a Primers should be stored at a concentration of 10–50 µM at −20°C.

2.3 Analysis of RT-PCR products by agarose gel electrophoresis

The results of RT-PCR can be analysed by agarose gel electrophoresis within an hour using a mini-gel apparatus. (see *Protocol 5*). The choice of gel concentration depends on the estimated size of the amplified DNA fragment: the range of separation in gels containing different concentrations of agarose is shown in *Table 1* and well-described elsewhere (5).

Table 1. Appropriate agarose concentrations for separating DNA fragments of various sizes

Agarose (%)	Linear DNA fragments (kb)
0.8	20–10
1.0	10–0.5
1.2	7–0.4
1.5	4–0.2

Protocol 5. Electrophoresis of PCR products in agarose gels

Equipment and reagents

- mini-gel electrophoresis apparatus
- agarose: low melting point agarose is recommended for separating DNA fragments less than 500 bp.
- 50 × TAE: 2 M Tris–acetate, 0.05 M EDTA
- loading dye (× 10): 100 µl 50 × TAE, 250 µl glycerol, 50 µl saturated bromophenol blue, 100 µl sterile distilled water
- ethidium bromide stock solution: dissolve 1 g of ethidium bromide in 100 ml of H_2O (10 mg/ml); wrap the container with aluminium foil

Method

1. Seal the plastic tray supplied with the mini-gel electrophoresis apparatus to form a mould. Set the mould on a horizontal section of the bench.

2. Add the appropriate amount of powdered agarose to a measured quantity of 1 × TAE buffer in an Ehrlenmeyer flask. Wrap the neck of the flask loosely with plastic wrap and heat the slurry in a microwave oven until the agarose dissolves.

3. Cool the agarose solution to about 60°C and add ethidium bromide stock solution to a final concentration of 0.5 µg/ml and mix well. Set the comb and pour the agarose solution.

4. After 30 min, remove the comb, and seal and mount the gel in the electrophoresis tank. Add 1 × TAE buffer to cover the gel to a depth of about 1 mm.

5. Take 18 µl of each PCR product and mix with 2 µl of loading dye. Carefully add the samples to the wells of the submerged gel with a micropipette. Run the gel until the bromophenol blue dye front has migrated the appropriate distance through the gel.

6. Examine the gel by ultraviolet light and take a photograph.

A representative result obtained using the procedure described in *Protocol 5* is shown in *Figure 1*. In this experiment, cDNA of the p53 tumour suppressor gene was amplified using primers with an external *Eco*RI site. In this way, the PCR product can be extracted from the agarose gel as described in *Protocol 6* and subcloned into the restriction site of a cloning vector (e.g. pBluescript; Stratagene) or a mammalian expression vector (pCDM8/neo; Invitrogen). These can then be expanded further in bacteria and used for the analysis of point mutations by dideoxynucleotide sequencing or transfection.

Protocol 6. Extraction of PCR products from agarose gels

Reagents

- buffer-saturated phenol, pH 8.0
- phenol:chloroform:isoamyl alcohol (25:24:1)
- 3 M sodium acetate, pH 5.2
- ice-cold absolute ethanol
- TE buffer: 10 mM Tris–HCl, pH 8.0, 1 mM EDTA

Method

1. Under a UV transilluminator cut out the band of interest from the agarose gel (see *Protocol 5*) and mince with a razor blade: transfer into a microcentrifuge tube.

2. Add 300 µl of buffer-saturated phenol and vortex well. Freeze the tube with dry-ice then thaw in a 37°C waterbath.

Protocol 6. *Continued*

3. Freeze and thaw the tube again.

4. Spin the tube at 15 000 *g* for 15 min at room temperature.

5. Collect the upper aqueous phase in another microcentrifuge tube.

6. Add the same volume of TE buffer to the lower phenol phase and freeze and thaw twice as in Steps 2–4.

7. Combine the two aqueous phases and extract with phenol:chloroform: isoamyl alcohol. Spin the tube at 15 000 *g* for 5 min at room temperature.

8. Collect the upper aqueous phase, add 1/10 volume of 3 M sodium acetate and 2 volumes of ice-cold absolute ethanol. Precipitate the DNA at −80°C for at least 2 h.

9. Spin at 15 000 *g* for 10 min and discard the liquid phase. Add 500 μl of 70% ethanol and wash the pellet using a vortex mixer.

10. Spin at 15 000 *g* for 10 min, discard the liquid phase and air-dry the pellet.

11. Re-suspend the DNA pellet with 10–20 μl of TE buffer.

The method described in *Protocol 6* is suitable for the recovery of DNA fragments larger than 500 bp. Various reliable DNA fragment isolation kits (*e.g.* GENECLEAN II, BIO-101; MERMAID, BIO-101; and Magic PCR Preps, Promega) are also available commercially.

3. Choice of primers and controls

3.1 Selection of primers

Choosing the appropriate primer set for PCR is essential for successful and specific amplification. First, the precise nucleotide sequence of the target cDNA must be known. Sources of sequences include original publications and DNA sequence databases such as GenBank (USA), EMBL (Europe) or DDBJ (Japan). Recently, these databases have been connected to each other and they can now be searched via telephone networks. For details of access to these databases, the following organizations can be contacted:

GenBank: Los Alamos National Laboratory
Los Alamos, NM 87545
USA
Telephone: +1–505–665–2177

EMBL: Postfach 10.2209
6900 Heidelberg
Germany
Telephone: +49–6221–387–258

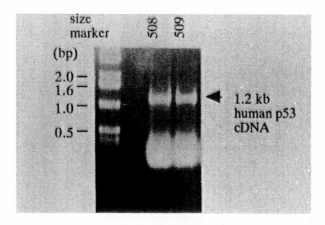

Figure 1. The result of 1% agarose gel electrophoresis (see *Protocol 5*) of RT-PCR amplified p53 cDNA from total RNA from gastric carcinoma tissue. Total RNA was prepared according to *Protocol 1*. RT-PCR was performed as described in *Protocols 3* and *4*. The primers used for PCR were 5'-AATTCAAAGTCTGTTCCGTCCCAGTAGATTTACTC-3' (sense) and 5'-GCGGAATTCAGTGGAGGATGTCAGTCTGAGTC-3' (anti-sense). These p53 fragments with external *Eco*RI sites were extracted from the agarose gel as described in *Protocol 6* and cloned into the *Eco*RI site of pBluescript. Deoxynucleotide sequencing revealed that case 509 had a valine to glutamine mutation at codon 272 of p53 tumour suppressor gene, while case 508 had wild-type p53 cDNA.

DDBJ: Laboratory of Genetic Information Analyis
Center for Genetic Information Research
National Institute of Genetics
1,111 Yata
Mishima, Shizuoka 411
Japan
Telephone: +81–559–75–0771

The appropriate PCR primers can be selected from cDNA sequences using computer software packages, such as OLIGO (National Biosciences Inc.), or public domain software, such as PRIMERS version 1.10 (by Greg Bristol at GREGB@lbes.medsch.ucla. edu.).

The specificity of each oligonucleotide can be determined by a database search using software based on the FastA algorism (6), such as the Genetics Computer Group sequence analysis software package (GCG, Inc., Madison, WI).

3.2 Choice of controls

Wang *et al.* (7) reported a plasmid, pAW109, which can generate positive control RNA for several genes (*Figure 2*). This plasmid contains 5' primers for 12 target genes (tumour necrosis factor (TNF), macrophage colony stimulating factor (M-CSF), platelet derived growth factor A (PDGF-A), PDGF-B,

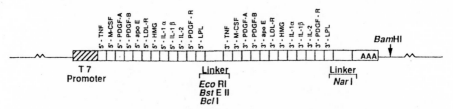

Figure 2. Structure of the pAW108 plasmid. The plasmid contains 5′ primers of 12 target genes connected in sequence followed by the complementary sequences of the 3′ primers in the same order. Restriction enzyme linkers are placed after the set of 5′ primers and after the set of 3′ primers to allow insertion of additional pairs as needed. The multiple primer region is flanked upstream by the T7 polymerase promoter and downstream by polyadenylated sequences (7).

apolipoprotein E (apoE) low density lipoprotein R (LDL-R), HMG, interleukin-1α (IL-1α), IL-1β, IL-2, PDGF-R, and lipoprotein lipase (LPL)) connected in sequence, followed by the complementary sequences of the 3′ primers in the same order. The primer sets for the amplification of each segment are shown in *Table 2*. This control RNA with a primer set to amplify the IL-1α fragment is provided in the GenAmp RNA PCR kit (Cetus). Amplification of this control RNA can be included to check the entire RT-PCR experiment.

4. Quantitative RT-PCR

Because of the high sensitivity of PCR, the RT-PCR system can detect trace amounts of mRNA which sometimes are difficult to detect by Northern blot analysis. For the rough estimation of the relative level of expression of target mRNA, one can include control reactions which amplify housekeeping genes, such as β-actin, and compare the ethidium bromide fluorescence intensity sample by sample. This method remains, however, only semi-quantitative. Wang *et al.* (7) have described a method for quantitative PCR analysis for mRNA. They constructed the pAW109 plasmid as described in Section 3.2 which can generate control RNA for PCR. There are two advantages to this control RNA. First, it can serve as an internal mRNA control for the RT reaction. Second, it can be used to generate a standard curve for quantitating the specific target mRNAs from experimental samples. RT-PCR was performed with sample RNA and different amounts of control RNA using ^{32}P end-labelled primers. After separating the PCR products in a polyacrylamide gel, the radioactivity of the target PCR product and control PCR product was determined by Çerenkov counting. Linear correlation was observed between radioactivity and the initial amount of control RNA.

Alhough the qualitative RT-PCR method of Wang *et al.* (7) used radioactive material, it can be applied to non-radioactive techniques, as reported

Table 2. Oligonucleotides of 5′ primers and 3′ primers of 12 target genes (7)

mRNA species	5′ primers	3′ primers	Size of PCR product. bp mRNA	cRNA
TNF	5′-CAGAGGGAAGAGTTCCCCAG-3′	5′-CCTTGGTCTGGTAGGAGACG-3′	325	301
M-CSF	5′-GAACAGTTGAAAGATCCAGTG-3′	5′-TCGGACGCAGGCCTTGTCATG-3′	171	302
PDGF-A	5′-CCTGCCCATTCGGAGGAAGAG-3′	5′-TTGGCCACCTTGACGCTGCG-3′	225	301
PDGF-B	5′-GAAGGAGCCTGGGTTCCCTG-3′	5′-TTTCTCACCTGGACAGGTCG-3′	217	300
apoE	5′-TTCCTGGCAGGATGCCAGGC-3′	5′-GGTCAGTTGTTCCTCCAGTTC-3′	270	301
LDL-R	5′-CAATGTCTCACCAAGCTCTG-3′	5′-TCTGTCTCGAGGGGTAGCTG-3′	258	301
HMG	5′-TACCATGTCAGGGGTACGTC-3′	5′-CAAGCCTAGAGACATAATCATC-3′	246	303
IL-1α	5′-GTCTCTGAATCAGAAATCCTTCTATC-3′	5′-CATGTCAAATTCACTGCTTCATCC-3′	420	308
Il-1β	5′-AAACAGATGAAGTCCTTCTCCAGG-3′	5′-TGGAGAACACCACTTGTTGCTCCA-3′	388	306
IL-2	5′-GAATGAATTAATAATTACAAGAATCCC-3′	5′-TGTTTCAGATCCCTTAGTTCCAG-3′	222	305
PDGF-R	5′-TGACCACCCAGCCATCCTTC-3′	5′-GAGGAGGTGTTGACTTCATTC-3′	228	300
LPL	5′-GAGATTTCTCTGTATGGCACC-3′	5′-CTGCAAATGAGACACTTTCTC-3′	277	300

cRNA = control RNA

by Powell and Kroon (8). Instead of end-labelling the primers, they included digoxigenin-11-dUTP in the same system. After the PCR product was separated by agarose gel electrophoresis and transferred to a nylon membrane, chemiluminescence was performed for quantitation. Each pair of signals corresponding to the target mRNA and the internal control was analysed by densitometry. The concentration of target mRNA was calculated from the relative sample and control peak areas (R) and the known number of molecules of control RNA added to the PCR reaction:

$$\text{mRNA molecules/mg RNA} = \frac{R \times (\text{copies of control RNA added})}{(\text{mg of sample RNA added}).}$$

The advantages of RT-PCR based quantitative mRNA analysis are as folows:

(a) High sensitivity: quantification of specific gene expression can be achieved from one microgram of total RNA.

(b) Rapidity by comparison with Northern blot analysis. The system described here can amplify target and control RNAs in one RT-PCR in a single tube without radioactive materials.

The disadvantages of this method are that good control RNA sources, such as pAW109, are required, as is optimization of the thermal cycle conditions to ensure that the amplification reaction remains linear.

References

1. Kawasaki, E. S., Clark, S. S., Coyne, M. Y., Smith, S. D., Champlin, R., Witte, O. N., and McCormick, F. P. (1988). *Proc. Natl Acad. Sci. USA*, **85**, 5698.
2. Almoguera, C., Shibata, D., Forrester, K., Martin, J., Arnheim, N., and Perucho, M. (1988). *Cell*, **3**, 549.
3. Rappolee, D., Wang, A., Mark, D., and Werb, Z. (1989). *J. Cell Biochem.*, **39**, 1.
4. Chirgwin, J. M., Przybyla, A. E., MacDonald, R. J., and Rugger, W. J. (1979). *Biochemistry*, **18**, 5294.
5. Sambrook, J., Fritsch, E. F., and Maniatis, T. (ed.) (1989). *Molecular cloning, a laboratory manual*, pp. 6.4–6.5. Cold Spring Harbor Laboratory Press, Cold Spring Harbor, NY.
6. Peason, W. R. and Lipman, D. J. (1988). *Proc. Natl Acad. Sci. USA*, **85**, 2444.
7. Wang, A. M., Doyle, M. V., and Mark, D. F. (1989). *Proc. Natl Acad. Sci. USA*, **86**, 9717.
8. Powell, E. E. and Kroon, P. A. (1992). *J. Lipid Res.*, **33**, 609.

A1

List of suppliers

American Type Tissue Culture Collection, 12301 Park Lawn Drive, Rockvilie, 20852, USA

Amersham International, Lincoln Place, Green End, Aylesbury, Bucks. HP20 2TP, UK.

Bio101, P.O. Box 2284, La Jolla, CA 92038, USA. Stratech, 61–63 Dudley Street, Luton, Beds. LU2 ONP, UK.

Biohit OY, Yerkkosaarenkatu 4, 00580, Helsinki, Finland. Croft-Biotech Ltd., Sanbeck Way, 25 Signet Court, Swann's Road, Cambridge CB5 8LA, UK.

Bio-rad Laboratories Ltd, Biorad House, Maylands Avenue, Hemel Hempstead, Herts. HP2 7TD, UK.

Boehringer–Mannheim UK (Diagnostics and Biochemicals) Ltd. Boehringer Mannheim Biochemicals, P.O. Box 50414, Indianapolis, IN 46250, USA. Bell Lane, Lewes, East Sussex BN7 1LG, UK.

C. A. Hendley (Essex) Ltd., Oakwood Hill Industrial Estate, Loughton, Essex, UK.

Cambridge Research Biochemicals, Gadbrook Park, Northwich, Cheshire CW9 7RA, UK.

Cetus–Perkin Elmer Ltd, Seer Green Support Centre, Beaconsfield, Bucks HP9 1QN, UK.

Clontech Laboratories Inc, 4030 Fabian Way, Palo Alto, CA 94303–9605, USA. Cambridge Bioscience, 25 Signet Court, Swann's Rd., Cambridge CB5 8LA, UK.

Dako Corporation, 6392 Via Reac, Carpinteria, CA 93013, USA. Dako Ltd, 16 Manor Courtyard, Hughenden Avenue, High Wycombe, Bucks. HP13 5RE, UK.

Digene Diagnostics Inc, 2301-B, Broadbirch Drive, Silver Spring, MD 20904, USA.

Enzo Diagnostics, 325 Hudson St., New York, NY 10013, USA. Cambridge Bioscience, 25 Signet Court, Swann's Rd., Cambridge CB5 8LA, UK.

Fluka Chemicals Ltd, Peakdale Road, Glossop, Derbyshire SK13 9XE, UK.

FMC Corporation, 5 Maple Street, Rockland, ME 04841–2994, USA. Flowgen Ltd, Broad Road, Sittingbourne, Kent ME9 8AQ, UK.

Gelman Sciences Ltd, Brackmills Business Park, Caswell Road, Northampton NN4 0EZ, UK.

Life Technologies Inc (Gibco BRL), 8400 Helgerman Court, Gaithersburg, MD 20877, USA. Life Technologies Ltd, Trident House, Renfrew, Paisley, PA3 4EF, UK.

Merck Ltd, Hunter Boulevard, Lutterworth, Leics., UK.

Millipore Corporation, 80 Ashby Road, Bedford, MA 01730. Millipore (UK) Ltd, The Boulevard, Blackmoor Lane, Watford, Herts. WD1 8YW, UK.

New England Biolabs, 32 Tozer Road, Beverly, MA 01915–5599, USA. 5'3' Inc. and CP Laboratories, P.O. Box 22, Bishops Stortford, Herts. CM23 3DX, UK.

Oncor, 209 Perry Parkway, Gaithersburg, MD 20877, USA. Alpha Laboratories, 40 Parham Drive, Eastleigh, Hants. SO5 4NU, UK.

Pharmacia-LKB Technology, Davy Avenue, Knowlhill, Milton Keynes MK5 8PH, UK.

Pierce Life Science Laboratories Ltd, Sedgewick Rd., Luton, Beds. LUU 9DT, UK.

Promega Corporation, 2800 Woods Hollow Road, Madison, WI 53711–5399, USA.

Promega Ltd, Delta House, Chilworth Research Centre, Southampton SO1 7NS, UK.

Scheicher and Schuell, P.O. Box 4-D-37582, Dassel, Germany. Anderman and Co. Ltd, 145 London Road, Kingston-upon-Thames, Surrey KTZ 6WH, UK.

Sigma Chemical Company Ltd, P.O. Box 14508, St Louis, MO 63178, USA. Fancy Road, Poole, Dorset BH17 7NH, UK.

Stratagene, Cambridge Innovation Centre, Unit 40, Cambridge Science Park, Cambridge CB4 4GF, UK.

Tropix, 47 Wiggins Avenue, Bedford, MA 01730, USA. NBS Biologicals, Edison House, 163 Dixons Hill Road, North Mymms, Hatfield, Herts. AL9 7JE, UK.

United States Biochemical USBTM, P.O. Box 22400, Cleveland, OH 44122, USA. Cambridge Bioscience, 25 Signet Court, Swann's Road, Cambridge CB5 8LA, UK.

Vector Laboratories Inc., 30 Ingold Road, Burlingame, CA 94010, USA. 16 Wulfric Square, Peterborough PE3 8RF, UK.

Zymed Laboratories Inc, 458 Carlton Court, South San Francisco, CA 94080, USA. Cambridge Bioscience, 25 Signet Court, Swann's Road, Cambridge CB5 8LA, UK.

Index

Index

Index

Index

Index

Printed in the United States
26879LVS00001B/140